Johannes Meyer

Light Field Methods for the Visual Inspection of Transparent Objects

Schriftenreihe Automatische Sichtprüfung und Bildverarbeitung
Band 15

Herausgeber: Prof. Dr.-Ing. habil. Jürgen Beyerer

Lehrstuhl für Interaktive Echtzeitsysteme
am Karlsruher Institut für Technologie

Fraunhofer-Institut für Optronik, Systemtechnik
und Bildauswertung IOSB

Light Field Methods for the Visual Inspection of Transparent Objects

by
Johannes Meyer

KIT Scientific Publishing

Karlsruher Institut für Technologie
Lehrstuhl für Interaktive Echtzeitsysteme

Light Field Methods for the Visual Inspection of Transparent Objects

Zur Erlangung des akademischen Grades eines Doktors der Ingenieurwissenschaften von der KIT-Fakultät für Informatik des Karlsruher Instituts für Technologie (KIT) genehmigte Dissertation

von M.Sc. Johannes Meyer

Tag der mündlichen Prüfung: 6. Dezember 2018
Referent: Prof. Dr.-Ing. habil. Jürgen Beyerer
Korreferent: Prof. Dr. Laurent Jacques

Impressum

KIT Scientific Publishing

Karlsruher Institut für Technologie (KIT)
KIT Scientific Publishing
Straße am Forum 2
D-76131 Karlsruhe

KIT Scientific Publishing is a registered trademark
of Karlsruhe Institute of Technology.
Reprint using the book cover is not allowed.

www.ksp.kit.edu

Print on Demand 2019 – Gedruckt auf FSC-zertifiziertem Papier

ISSN 1866-5934
ISBN 978-3-7315-0912-7
DOI 10.5445/KSP/1000091872

Acknowledgments

I would like to express my gratitude to Professor Dr.-Ing. Jürgen Beyerer for providing me the opportunity to work under his guidance at the Vision and Fusion Laboratory of the Karlsruhe Institute of Technology (KIT, Germany) in cooperation with the Fraunhofer-Institute of Optronics, System Technologies and Image Exploitation (IOSB, Germany) and for constantly supervising and supporting my research. Many thanks also go to Professor Dr. Laurent Jacques from the Université Catholique de Louvain (UCLouvain, Belgium) for serving as the second reviewer and for his valuable comments and suggestions regarding my thesis. A big thank is addressed to the department Visual Inspection of the IOSB, especially to Matthias Hartrumpf for boosting my experiments by realizing my ideas as extensions of his laser scanners. I am also very grateful for the thorough proofreading of my thesis by my colleague Christian Frese.

Furthermore, I want to thank my loving girlfriend Ellen Apolinar for unremittingly listening to the rehearsals of my talks, for her constructive feedback and for reminding me to work on my thesis instead of playing computer games. Also, my family, in particular my parents Marion and Thomas Meyer, constantly supported me with encouraging words and delicious meals during my home stays. Finally, I am very happy about my two good friends Dirk Gutting and Joscha Borné, which are reliable companions for me since the beginning of my time in Karlsruhe.

Karlsruhe, December 2018 *Johannes Meyer*

Kurzfassung

Objekte aus transparenten Materialien sind im menschlichen Alltag allgegenwärtig. Sie werden verwendet zur Fertigung von Windschutzscheiben für Autos und Flugzeuge, Scheiben zur Abdeckung von Scheinwerfern in Automobilen, Lebensmittelverpackungen, hochpräzisen optischen Instrumenten und von Plastiklinsen zur Strahlformung in lasergestützten Augenoperationen. Bei letzterem Beispiel wird deutlich, dass solche Objekte höchsten Qualitätsansprüchen genügen müssen. Je nach Anwendung dürfen transparente Objekte keinerlei Materialdefekte aufweisen, wie bspw. lichtabsorbierende Fremdkörper (z.B. Staub- oder Schmutzpartikel), lichtstreuende Verunreinigungen (z.B. eingeschlossene Luftblasen) oder Oberflächenkratzer. Um sicherzustellen, dass ein produziertes transparentes Objekt die Qualitätsanforderungen für eine gewisse Anwendung erfüllt, muss es einer Sichtprüfung unterzogen werden.

Menschliche Sichtprüfer stellen eine Möglichkeit der Qualitätskontrolle dar. Diese Arbeit ist für den Menschen jedoch sehr ermüdend und die Ergebnisse sind oft von der subjektiven Entscheidung des Sichtprüfers beeinflusst. Hinzu kommt die Gefahr übersehener Materialdefekte, was je nach Anwendungsfall gravierende Folgen haben kann.

Automatische Sichtprüfsysteme stellen eine Alternative zur manuellen Sichtprüfung dar. Solche Systeme arbeiten 24 Stunden pro Tag und liefern zuverlässige und objektive Prüfentscheidungen. Obwohl ausgereifte Systeme zur Prüfung von Flachglas oder anderen einfach geformten Prüfobjekten weit verbreitet sind, stellt die automatische Sichtprüfung komplex geformter transparenter Objekte weiterhin eine Herausforderung mit diversen ungelösten Forschungsfragen dar. Die optischen Komponenten dieser Systeme sind meist maßgeschneiderte Lösungen für ein spezielles Prüfobjekt. Die notwendigen Designprozesse sind entsprechend langwierig und setzen ein hohes Maß an Expertenwissen voraus. Ein solches System schnell an einen neuen Prüflingstyp anzupassen ist meist unmöglich.

Transparente Objekte und darin enthaltene Materialdefekte beeinflussen die Ausbreitungsrichtung von sie durchdringendem Licht. Zur visuellen Inspektion betrachten Menschen transparente Objekte daher meistens aus mehreren Blickwinkeln, da insbesondere lichtstreuende Defekte nur aus bestimmten Richtungen sichtbar sind. Sie erfassen sozusagen Lichtfelder, die sowohl Informationen über den Ort als auch über die Ausbreitungsrichtung von durch den Prüfling transmittierten Lichtstrahlen beinhalten. Diese Beobachtung wird in der vorliegenden Arbeit aufgegriffen. Für alle grundlegenden Bestandteile eines Sichtprüfsystems, der Beleuchtungskomponente, der Bildaufnahmeeinheit und der Bildauswertung, werden neue Methoden und Verfahren basierend auf dem Lichtfeldkonzept vorgestellt. Im Folgenden sind die Hauptbeiträge dieser Arbeit zusammengefasst.

Zur Aufnahme von Lichtfeldern von transparenten Objekten wird ein neuartiges Sensorkonzept vorgestellt. Der sogenannte Laser Deflection Scanner kann Lichtfelder mit hoher Orts- und Winkelauflösung erfassen. Zusätzlich wird ein Gradientenbegriff für vierdimensionale Lichtfelder formuliert, der auf entsprechenden Distanzfunktionen beruht. Mit Hilfe dieses Gradienten können Unstetigkeiten der Ausbreitungsrichtung des Lichts zwischen benachbarten Objektpunkten detektiert werden. Die Generalisierte Cramèr-von Mises-Distanz wird angepasst, um als passende Distanzfunktion zu fungieren. Mit einem neuen Algorithmus kann zudem die Komplexität der Berechnung dieser Distanz drastisch reduziert werden. Die Kombination aus Laser Deflection Scanner und dem Lichtfeldgradienten ermöglicht eine schnelle, für industrielle Anwendungen geeignete Inspektion komplex geformter transparenter Objekte.

Des Weiteren werden im Rahmen dieser Arbeit zwei Algorithmen zur Bildfusion eingeführt, die unter variablen Beleuchtungsbedingungen aufgenommene Kamerabilder eines konventionellen Inspektionssystems fusionieren und dabei sämtliche Sollstrukturen des Prüflings unterdrücken und Defekte hervorheben. Dazu wird Wissen über den Prüfling und das Prüfsystem ausgenutzt, das zuvor mittels automatischer Lernverfahren gelernt wurde. Durch dieses Verfahren wird die zur Anpassung des Systems an neue Prüflingstypen nötige Expertise und Zeit stark reduziert und die Qualität der resultierenden Inspektionsbilder deutlich verbessert.

Es wird ein Ansatz zur inversen Lichtfeldbeleuchtung vorgestellt, anhand dessen ein transparentes Objekt mit nur einer Bildaufnahme komplett inspiziert werden kann. Während der Inspektion wird dazu der Prüfling mit einem inversen Lichtfeld bestrahlt, das sämtliche Brechungseffekte im Prüfling invertiert. Alle defektfreien Bereiche des Prüflings transformieren das inzidente Lichtfeld in Strahlen, die parallel zur optischen Achse einer telezentrischen Kamera verlaufen und von dieser somit eingefangen werden. Vorhandene Defekte lenken das Licht ab, was zu dunklen Stellen im Bild führt und die Defekte sichtbar macht. Zur Emittierung des inversen Lichtfeldes wird das Konzept eines Lichtfeldgenerators vorgestellt und eine prototypische Umsetzung gezeigt.

Die vorgestellten Verfahren werden mittels simulierter und realer Experimente umfassend evaluiert. Die Ergebnisse zeigen, dass sich alle Methoden für den Einsatz zur automatischen Sichtprüfung transparenter Objekte eignen und dass sie dem Stand der Wissenschaft und Technik in diversen Aspekten überlegen sind.

Abstract

Objects made out of transparent materials play crucial roles in humans' everyday life. They are employed as windshields for automobiles or airplanes, as cover glasses for automotive headlamps, as transparent containers for food or liquids, to build high precision optical instruments or even as plastic lenses to guide laser beams in an eye operation. Especially when considering the latter example it is obvious, that such transparent objects have to meet high quality requirements. Depending on the application, transparent objects have to be free from material defects, such as light absorbing contaminants (e.g., dirt particles), light scattering impurities (e.g., air bubbles) or surface scratches and so on. To ensure that produced transparent parts are of the quality required for the intended application, they have to be visually inspected.

On the one hand, the visual inspection of transparent objects can be carried out by human visual inspectors. However, human visual inspection represents a fatiguing task which is not very robust and prone to subjective results or even to unrevealed defects what can have severe consequences.

On the other hand, automated visual inspection systems can be employed for the visual inspection. Such systems can operate for 24 hours a day and yield reliable and objective results. Although there exist elaborated machine vision systems for inspecting flat glass plates or other objects with a simple geometry, the automated visual inspection of complex-shaped transparent objects still represents a challenging task with several open research questions. The optical components of existing solutions are usually tailored specifically to a certain type of test object to obtain an optimal inspection image. This involves many tedious and time-consuming empirical experiments by an expert. Quickly adapting such systems to a new type of test object is impractical.

A transparent object itself and the material defects have an effect on the direction of propagation of the transmitted light. This is why when inspecting a transparent object, humans usually look at the object from different perspectives. In other words, they acquire light fields which contain spatial and angular information about the transmitted light. Especially scattering material defects might only be visible under a certain angle. According to this observation, this thesis introduces methods based on the concept of light fields for all main components of a visual inspection system, the illumination source, the sensor device and the signal processing algorithms. The major contributions of this thesis are summarized in the following.

A novel sensor device, a laser deflection scanner, is presented that acquires light fields of transparent objects with a high spatial and a high angular resolution. By means of suitable distance functions, a gradient is formulated for light fields that allows to detect discontinuities of the light's direction of propagation between spatially adjacent object points. The Generalized Cramér-von Mises Distance is adapted to serve as an adequate distance function. An algorithm is introduced that greatly reduces the computational complexity of this distance. This allows to reveal material defects even in complex-shaped objects in an inspection time feasible for industrial production lines.

Furthermore, this thesis introduces two image fusion algorithms that combine the images obtained with a conventional inspection system to an inspection image in which all intended object structures are suppressed and material defects are shown with high contrast.

For this purpose, the algorithms exploit information about the test object and the inspection system which can be acquired via an automated learning procedure. The amount of expertise and time to adapt the inspection system to a new type of test object is reduced and the quality of the resulting inspection images is increased.

An inverse light field illumination method is presented that inverts the refraction effects evoked by a transparent test object. During the inspection, the test object is observed with a telecentric camera. All defect-free parts of the object will transform the incident light rays into rays propagating parallel to the optical axis which will be captured by the camera. Present material defects will be revealed since they deflect the transmitted light rays leading to dark regions in the camera image. With this approach, a test object can be inspected by acquiring a single image only. To emit the inverse light field, the optical concept of a light field generator is described and a first prototype is presented.

A thorough evaluation is performed by means of simulated and real experiments. The results prove the suitability of all the introduced methods for the visual inspection of transparent objects and state their superiority over the state of the art with respect to several criteria.

Contents

List of symbols

Conventions

$\mathrm{a}, \mathrm{b}, \ldots$	identifiers (roman, lower-case or
$\mathrm{A}, \mathrm{B}, \ldots$	capital letters)
$x, y, \ldots$	scalar, continuous, real-valued variables (italic,
$A, B, \ldots$	lower-case letters, sometimes capital letters)
$\mathbf{x}, \mathbf{y}, \ldots$	vectors or points of continuous values (bold, lower-case letters)
$\mathbf{x}^\mathsf{T}$	transposed vector
$\mathbf{A}, \mathbf{B}, \ldots$	matrices (bold, capital letters)
$\mathcal{A}, \mathcal{B}, \ldots$	sets (calligraphic, capital letters)
$M, N, \ldots$	number of elements of a set (italic, capital letters)
$i, j, \ldots$	integer variables (italic, lower-case letters)
$\mathbf{j}, \mathbf{m}, \mathbf{u}$	vectors of integers (bold, lower-case letters)
$\mathtt{a}, \mathtt{b}, \ldots$	random variables (roman, typewriter font)
$\boldsymbol{\mathtt{a}}, \boldsymbol{\mathtt{b}}, \ldots$	random vectors (bold, typewriter font)
$\mathfrak{f}, \mathfrak{g}, \ldots$	summed area tables (fraktur, lower-case letters)

Symbols

$\mathbf{0}$	zero vector, $\mathbf{0} = (0, \ldots, 0)^\mathsf{T}$
α, θ, φ	continuous angles
Δ	distance of a test object to a light source

For reasons of clarity, symbols which are only used once throughout this thesis or which are of minor importance, are omitted in this list. In the case of symbols having multiple similar variants, e.g., $\mathbf{p}^\mathrm{a}, \mathbf{p}^\mathrm{b}$, only the base type, e.g., $\mathbf{p}$, is shown here.

δ_α	spatial displacement on the focal plane of a lens to the optical axis corresponding to an incident light bundle with an angle of α with respect to the optical axis
$\gamma(k,l)$	amount of probability mass moved from bin k of h_1 to bin l of h_2 in the calculation of $\mathrm{EMD}(h_1,h_2)$
λ	wavelength of the light
$\hat{\mu}$	estimated mean value of an image region
$\boldsymbol{\mu}(\mathbf{x})$	vector of mean gray values of a series of multi-channel images, $\boldsymbol{\mu}(\mathbf{x}) = (\mu_1(\mathbf{x}),\ldots,\mu_C(\mathbf{x}))^\intercal$
$\hat{\boldsymbol{\mu}}(\mathbf{x})$	estimation of $\boldsymbol{\mu}(\mathbf{x})$
$\boldsymbol{\xi}$	vector of parameters
ρ	the plenoptic function $\rho(x,y,z,\theta,\varphi,\lambda,\tau)$
$\hat{\sigma}$	estimated standard deviation of an image region
τ	time
$\theta_\Delta,\varphi_\Delta$	angular pitches between discrete angles that can be emitted by a light field generator
$\theta_\mathrm{t},\varphi_\mathrm{t}$	total ranges of angles that can be emitted by a light field generator
Ω	domain of a function
$a(\mathbf{m},\mathbf{j})$	deflection map
$\widetilde{a}(\mathbf{m},\mathbf{j})$	normalized deflection map
A	index of the first element in a correspondence vector that is greater than a specified threshold
$A_\mathrm{h},A_\mathrm{v}$	angular resolution of a light field generator in the horizontal and the vertical direction
b	image distance
$\mathbf{b}$	vector of the sizes of the integration kernel used by the localized cumulative distribution
$\mathrm{bin}(u)_i$	i-th bit of the binary representation of u beginning at the most significant bit

$c(k, l)$	costs of moving one unit probability mass from bin k of h_1 to bin l of h_2 in the calculation of EMD(h_1, h_2)
$\mathbf{c}^{\mathrm{ML}}$	the center of a microlens in the coordinates of the plane of the microlens array, $\mathbf{c}^{\mathrm{ML}} = \left(c_x^{\mathrm{ML}}, c_y^{\mathrm{ML}}\right)^{\mathsf{T}}$
$\mathbf{c}(m, n)$	correspondence vector of a light transport matrix for camera pixel (m, n)
C	number of channels of a multi-channel image
CMD(f, g)	Generalized Cramér-von Mises Distance between two probability density functions f and g
CNR	(peak) contrast-to-noise ratio
d	diameter of a lens
$\mathbf{d}$	direction of a ray of sight, $\mathbf{d} = (d_x, d_y, d_z)^{\mathsf{T}}$
$e_{\mathrm{h}}, e_{\mathrm{v}}$	pitches of pixel elements in the horizontal and the vertical direction
E	index of the last element in a correspondence vector that is greater than a specified threshold
EMD(h_1, h_2)	Earth Mover's Distance between two histograms h_1 and h_2
f	focal length of a lens
$f(\cdot)$	arbitrary function
$f(\mathbf{x})$	probability density function
$F(\mathbf{x}, \mathbf{b})$	localized cumulative distribution evaluated at position $\mathbf{x}$ and with kernel sizes $\mathbf{b}$
g	object distance
$g(\mathbf{x})$	scalar image
$\mathbf{g}(\mathbf{x})$	multi-channel image
h	histogram
$i(m, n)$	a two-dimensional array of scalar values
$\mathscr{I}(\mathbf{g})$	information contained in $\mathbf{g}$ about a test object, relevant for its visual inspection

$\mathbf{j}$	discrete coordinates with an angular interpretation, $\mathbf{j} = (j, k)^\intercal$
$\mathbf{J}$	Jacobian matrix
J, K	size of the angular domain of a deflection map in the horizontal and the vertical direction
L	light field incident to a test object
L'	light field exiting a test object
$\widetilde{L}$	light field inverse to L
$\mathcal{L}$	set of coordinates corresponding to the pixels of a light field generator
$\mathbf{m}$	discrete (image) coordinates, $\mathbf{m} = (m, n)^\intercal$
M, N	number of elements in a set, number of bins in a histogram or number of pixels of an image in the horizontal and the vertical direction
$\mathcal{M}$	set of valid shifts of probability mass in the calculation of the EMD
$\mathbf{n}$	normal vector of a surface, $\mathbf{n} = (n_x, n_y, n_z)^\intercal$
$\mathbf{n}(\mathbf{x})$	unwanted (noise) component of a multi-channel signal, $\mathbf{n}(\mathbf{x}) = (n_1(\mathbf{x}), \ldots, n_C(\mathbf{x})^\intercal$
$\mathbb{N}$	set of natural numbers
$\mathbf{o}$	origin of a ray of sight, $\mathbf{o} = (o_x, o_y, o_z)^\intercal$
$\mathcal{O}(n)$	function class of asymptotic upper bounds (Big O notation)
$\mathbf{p}$	arbitrary point, $\mathbf{p} = (p_x, p_y, p_z)^\intercal$
$P(\cdot)$	probability
$\mathbf{P}^{\mathrm{noise}}(\mathbf{x})$	power of the noise component of a multi-channel signal, $\mathbf{P}^{\mathrm{noise}}(\mathbf{x}) = (P_1^{\mathrm{noise}}(\mathbf{x}), \ldots, P_C^{\mathrm{noise}}(\mathbf{x}))^\intercal$
$\mathbf{P}^{\mathrm{signal}}(\mathbf{x})$	power of the signal component of a multi-channel signal, $\mathbf{P}^{\mathrm{signal}}(\mathbf{x}) = (P_1^{\mathrm{signal}}(\mathbf{x}), \ldots, P_C^{\mathrm{signal}}(\mathbf{x}))^\intercal$
Q	number of light sources in a scene
$\mathcal{Q}$	set of quantized image values

r	factor specifying a defect's size with respect to the area of a camera pixel
R	radiance of a light source w.r.t. a given direction
R'	radiance of a traced ray of sight
$\mathbb{R}$	set of real-valued numbers
$\mathbb{R}_+$	set of positive real-valued numbers including 0
$\tilde{s}$	maximum image value of the image of a defect
$\mathfrak{s}(m,n)$	a summed area table
$\mathrm{sc}(m,n)$	feature scatter count at position (m,n)
$s_\mathrm{h}, s_\mathrm{v}$	pitches of the microlenses in a microlens array in the horizontal and the vertical direction
$\mathrm{sw}(m,n)$	feature scatter width at position (m,n)
$\mathbf{s}(\mathbf{x})$	wanted (signal) component of a multi-channel signal, $\mathbf{s}(\mathbf{x}) = (s_1(\mathbf{x}), \ldots, s_C(\mathbf{x}))^\mathsf{T}$
$\hat{\mathbf{s}}(\mathbf{x})$	estimation of $\mathbf{s}(\mathbf{x})$
S	number of detector elements in a scene
$\mathbf{S}_m, \mathbf{S}_n$	filter kernels of the Sobel operator in m- and n-direction
$\mathrm{SNR}_c(\mathbf{x})$	signal-to-noise ratio of channel c of a multi-channel signal
$\mathbf{SNR}(\mathbf{x})$	signal-to-noise ratio of a multi-channel signal, $\mathbf{SNR}(\mathbf{x}) = (\mathrm{SNR}_1(\mathbf{x}), \ldots, \mathrm{SNR}_C(\mathbf{x}))^\mathsf{T}$
$\widehat{\mathbf{SNR}}(\mathbf{x})$	estimation of $\mathbf{SNR}(\mathbf{x})$
$\mathcal{S}$	set of support positions
t	threshold value
$\mathbf{T}$	light transport matrix
$\mathcal{T}_\mathrm{n}$	training set of images corresponding to the noise component of an image formation process
$\mathbf{u} = (u,v)^\mathsf{T}$	discrete coordinates corresponding to the pixels of a programmable light source

$\widetilde{\mathbf{u}}$	central pixel under a microlens of a light field generator, $\widetilde{\mathbf{u}} = (\widetilde{u}, \widetilde{v})^\mathsf{T}$
$\mathcal{U}(\mathbf{x})$	set of coordinates arranged around $\mathbf{x}$
U, V	number of pixels of a programmable light source in the horizontal and the vertical direction
$\mathbf{v}(\mathbf{x})$	indicator function for the noise-based selective channel fusion, $\mathbf{v}(\mathbf{x}) = (v_1(\mathbf{x}), \ldots, v_C(\mathbf{x}))^\mathsf{T}$
V	magnification of an optical imaging system
$\mathbf{w}(\mathbf{m})$	mean deflection direction at measurement point $\mathbf{m}$ in x- and y-direction, $\mathbf{w}(\mathbf{m}) = (w_x(\mathbf{m}), w_y(\mathbf{m}))^\mathsf{T}$
$\mathbf{x}$	continuous coordinates, $\mathbf{x} = (x, y)^\mathsf{T}$ or $\mathbf{x} = (x, y, z)^\mathsf{T}$ depending on use-case
$(x^\mathrm{b}, y^\mathrm{b})^\mathsf{T}$	continuous image coordinates
$(x^\mathrm{c}, y^\mathrm{c}, z^\mathrm{c})^\mathsf{T}$	continuous camera coordinates
$(x^\mathrm{w}, y^\mathrm{w}, z^\mathrm{w})^\mathsf{T}$	continuous world coordinates
$\mathbb{Z}$	set of integers

Operators

$\lvert \cdot \rvert$	absolute value of a scalar variable
$\lVert \cdot \rVert$	norm of a vector
$\lVert \cdot \rVert_\mathrm{F}$	Frobenius norm of a matrix
$\frac{\partial g}{\partial x}$	partial derivative of the function g with respect to the variable x
$d\{\cdot, \cdot\}$	distance function for deflection maps
$d_\mathrm{E}\{\cdot, \cdot\}$	Euclidean distance function for deflection maps
$\operatorname{grad} g(\mathbf{x})$	gradient of the function $g(\mathbf{x})$
$\operatorname{Interpol}\{\cdot\}$	interpolation operator

Acronyms

CAD	computer aided design
CMD	Generalized Cramér-von Mises Distance
CNR	(peak) contrast-to-noise ratio
CPU	central processing unit
CSG	constructive solid geometry
DIN	Deutsches Institut für Normung
DMD	digital micromirror device
DSLR	digital single-lens reflex
EMD	Earth Mover's Distance
FPGA	field programmable gate array
ISO	International Organization for Standardization
LCD	localized cumulative distribution
	liquid crystal display
LCoS	liquid crystal on silicon
LED	light emitting diode
OVID	optical variable image device
PBRT	physically based renderer
PCB	printed circuit board
PSD	position sensitive detector
RAM	random-access memory
RGB	red, green and blue color coordinate
SAT	summed area table
SNR	signal-to-noise ratio

TCP	transmission control protocol
USB	universal serial bus
WLAN	wireless local area network

1 Introduction

Objects made of transparent materials, such as glass or certain plastics, are of the utmost importance for humans' everyday life. They are employed, e.g., in house construction as windows or even as more substantial building parts, such as transparent walls. Transparent glass or plastic covers are used to protect the sensitive displays of smartphones or other electronic devices and therefore have to be both very durable and light. Besides that, transparent materials are also employed to produce containers for food or pharmaceuticals so that the content is clearly visible while being properly sealed at the same time. Further fields of applications of transparent objects are in the automotive and aircraft industries. The windshields of automobiles and aircraft have to protect the driver or the pilot from the surrounding environment without impairing their view. Furthermore, the lenses and cover glasses of a car's headlights are made out of transparent materials. High precision optical elements which are, e.g., used in scientific experiments or to build optical sensing devices, represent an important application of transparent objects as well. Such components are also employed in medical technology, e.g., as so-called personal interfaces that guide laser beams to the intended positions during an eye operation.

Depending on the application, transparent objects have to meet different quality requirements. For example, glass windows only have to be free from those contaminants that can be seen by a human looking through them, but their index of refraction is allowed to vary over a large interval. In contrast, the personal interfaces used in eye surgery must not contain any absorbing or scattering contaminants that could scatter the incident laser light what could result in severe injury to the patient. The index of refraction can be required to vary by no more than 10^{-6} for the optical elements employed in high precision optical devices [Int96]. In order to ensure that the quality of a transparent object is suitable for its intended purpose, both objective specifications of the quality requirements and a way of evaluating these requirements are necessary.

Specification of the quality of transparent objects The norm DIN ISO 10110 defines a standard for describing the important properties of transparent objects [Int96]. Among other things, some sections of this norm also state how to define the quality requirements of a transparent object with respect to different types of material defects:

- **Section 2:** Imperfections leading to stress birefringence.
- **Section 3:** Enclosed bubbles and other inclusions.
- **Section 4:** Inhomogeneities (so-called striae) of the distribution of the index of refraction.
- **Section 5:** Differences regarding the intended three-dimensional shape.
- **Section 7:** Local surface imperfections (e.g., scratches).
- **Section 8:** Imperfections affecting the texture of the object's whole surface (e.g., the surface roughness).

In order to evaluate whether a given test object meets the requirements specified according to one or more sections of DIN ISO 10110, the necessary information of all occurrences of material defects of the corresponding type in the test object have to be obtained. For most of the definitions of DIN ISO 10110, the respective defects have to be detected, located, and measured in size. For example, for the evaluation of a specification according to Section 3 of DIN ISO 10110, the locations and sizes of all enclosed bubbles and inclusions having a size within a certain range have to be known. Section DIN ISO 10110-3 defines a relation that allows to compare the severity of one larger inclusion or bubble with a local accumulation of multiple small occurrences.

There are mainly two possibilities for performing the visual inspection of a transparent object, i.e., acquiring the required information about any material defects. On the one hand, human workers can carry out this task, and on the other hand, automated visual inspections can be employed. For humans, visual inspection in general is a monotonous, laborious, and fatiguing task, which can lead to subjective results which are not reproducible [Sch73]. Additionally, the documentation of the inspection results can be costly. Due to the slow speed of such inspection and high labor costs, it is very expensive. Manual visual inspection of transparent parts is especially uncomfortable, since they have to be observed in front of very bright light sources in order to see every defect. Reflections of the strong illumination directly into the eyes of the human can cause temporary blindness or reduced inspection accuracy. In turn, this can result in overlooked defects, which might have disastrous consequences, depending on the application. An adequately designed and developed automated visual inspection system can work for 24 hours a day without fatigue, yield reproducible results, and enable a detailed and comprehensive documentation.

As stated before, in order to automatically assess the quality of a transparent object, e.g., with respect to specifications according to DIN ISO 10110, it is necessary to visualize and measure material defects of interest. Considering the state of the art with respect to the visual inspection of transparent objects (see Chapter 2), absorbing contaminants or scattering structures (e.g., scratches, enclosed bubbles), i.e., Sections 3 and 7 of DIN ISO 10110, represent the most important defects that have to be detected. For some types of test objects and certain kinds of defects, there are approved and robust visual inspection systems (see Section 2.2). However, some inspection tasks, e.g., the visualization of scattering defects inside transparent test objects with a complex three-dimensional geometry, still constitute an open issue with several research questions that have to be answered. The following section systematically describes the problem addressed in this thesis.

1.1 Automated visual inspection of transparent objects

A typical visual inspection system for transparent objects consists of three main components: one or more illumination sources, one or more sensor devices and an image processing unit. Figure 1.1 illustrates the interaction of the components. The illumination unit emits a light field L (see Section 1.2), which can be controlled by the image processing unit. For example, a monitor, i.e., a spatially programmable light source, could serve as the illumination and the displayed image would be controlled by the image processing unit. By means of different physical effects, the transparent test object modulates the light field L, resulting in

$$L' = f_{\mathrm{phys}}\left(L, \boldsymbol{\xi}_{\mathrm{geom}}\right) , \tag{1.1}$$

with the function f_{phys} representing the modulation and $\boldsymbol{\xi}_{\mathrm{geom}}$ denoting the involved and manipulable parameters.

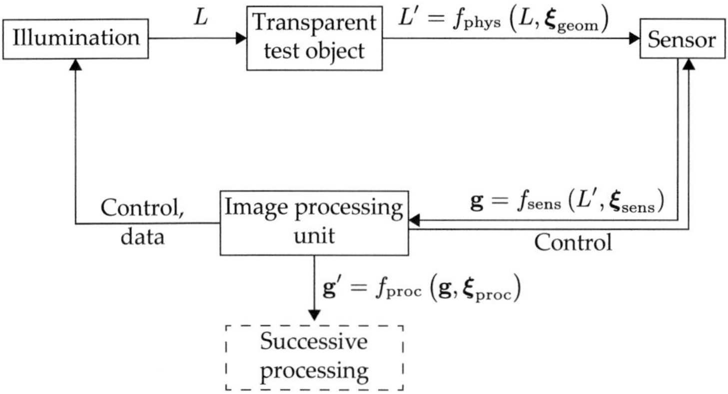

Figure 1.1: Main components of a visual inspection system for transparent objects: the illumination components emit a light field L that is modulated by the transparent test object. The modulated light field L' is captured by a sensor device which transfers the corresponding data to an image processing unit that also controls both the illumination and the sensor.

Since all optical elements affecting L' are assumed to be included in the illumination unit, $\boldsymbol{\xi}_{\mathrm{geom}}$ describes the geometric relations of the test object and L, i.e., their relative alignment and orientation. Thus, information about the test object is encoded in the modulated light field L'. A sensor captures a section of L' and converts it into a digital image

$$\mathbf{g} = f_{\mathrm{sens}}\left(L', \boldsymbol{\xi}_{\mathrm{sens}}\right) , \tag{1.2}$$

with the function f_{sens} denoting the acquisition process, which depends on L' and the parameters $\boldsymbol{\xi}_{\mathrm{sens}}$. The vector $\boldsymbol{\xi}_{\mathrm{sens}}$ contains geometric information about the alignment and orientation of the sensor but also about the acquisition parameters, e.g., the exposure time or amplification factors. Since the resulting image $\mathbf{g}$ could consist of multiple channels, it is expressed as a vector. An image processing unit, e.g., a computer running a suitable software, processes the image g in order to extract those components which are relevant for the inspection task and to suppress noise or other irrelevant parts. The image processing can be described via

$$\mathbf{g}' = f_{\text{proc}}\left(\mathbf{g}, \boldsymbol{\xi}_{\text{proc}}\right) , \tag{1.3}$$

with f_{proc} denoting the image processing algorithms and $\boldsymbol{\xi}_{\text{proc}}$ their corresponding parameters. In order to allow successive processing steps, e.g., classification algorithms, to determine whether the inspected test object is flawless or not, the relevant information $\mathscr{I}(\mathbf{g})$ contained in g about the test object has to be maximized.

By means of this representation, the task of designing an inspection system for a given test object type can be formalized as the following optimization problem:

$$\underset{\boldsymbol{\xi}}{\operatorname{argmax}}\, \mathscr{I}(\mathbf{g}') \quad \text{s.t.} \quad \boldsymbol{\xi} \text{ physically feasible} , \tag{1.4}$$

with $\boldsymbol{\xi} = \left(L, \boldsymbol{\xi}_{\text{geom}}^{\mathsf{T}}, \boldsymbol{\xi}_{\text{sens}}^{\mathsf{T}}, \boldsymbol{\xi}_{\text{proc}}^{\mathsf{T}}\right)^{\mathsf{T}}$ denoting the concatenation of all the parameters. Of course, there are additional properties of a visual inspection system (see Section 1.1.2) which represent a secondary optimization target.

This optimization problem is usually solved by engineers in a semi-automated and iterative way: An initial approach for the hardware setup is determined based on the engineer's knowledge and on the requirements of the inspection task. Then, image processing algorithms are chosen and parametrized and their suitability for detecting the relevant material defects is evaluated. Depending on these results, the hardware setup or the algorithms are improved. These steps are repeated until the whole system fulfills the inspection requirements or no more improvements are possible. However, the optimization problem specified in Eq. (1.4) is an ill-posed problem since its solution does not change continuously with the initial conditions, particularly with respect to the test object's geometry (cf. point 5 robustness in Section 1.1.2). Depending on the manufacturing process, the test object's production can lead to variations of its three-dimensional shape that would each require a different parameter vector $\boldsymbol{\xi}$ in order to obtain an optimal inspection system (see Fig. 1.2). There are types of test objects, the so-called cooperative test objects, for which the variations do not require separate solutions and the design practice described above can lead to valid results. However, for uncooperative test objects, this approach is likely to fail.

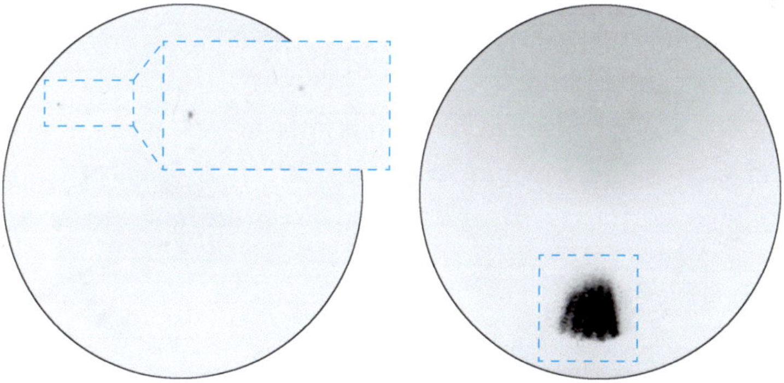

(a) (b)

Figure 1.2: The influence of geometric variations on inspection images of cylindrically shaped preforms made out of glass, so-called gobs: (a) for a test object that is not affected by geometric variations, the inspection image shows only the material defects (two particle inclusions inside the blue rectangle); (b) geometric variations result in strong reflections of the light sources so that the highlighted region can not be completely inspected.

1.1.1 'Cooperative' and 'uncooperative' transparent test objects

A cooperative transparent test object has several qualitative properties that ease the design process of a visual inspection system:

- Its surface is smooth and unstructured,
- it has a low level of geometric complexity, i.e., only few edges, and its distinct surfaces are mostly flat, and
- the variations of the test object's properties do not exceed the range for which a determined ξ leads to a near optimal inspection setup.

Any test object not satisfying all of these conditions is referred to as an uncooperative test object.

Examples of cooperative test objects include glass plates or planar cover glasses. All types of lenses, windshields, glass bottles, and other objects with a more sophisticated design are uncooperative test objects. Figure 1.3 shows examples of the two different types of test objects.

1.1.2 Evaluation criteria for automated visual inspection systems

If an automated visual inspection system is to be installed to solve an inspection task in a typical industrial application, potential inspection systems are usually evaluated according to the following criteria [Chi82; Mal03]:

1. **Versatility:** The set of different types of material defects (see page 2) that can be simultaneously detected (the more, the better).

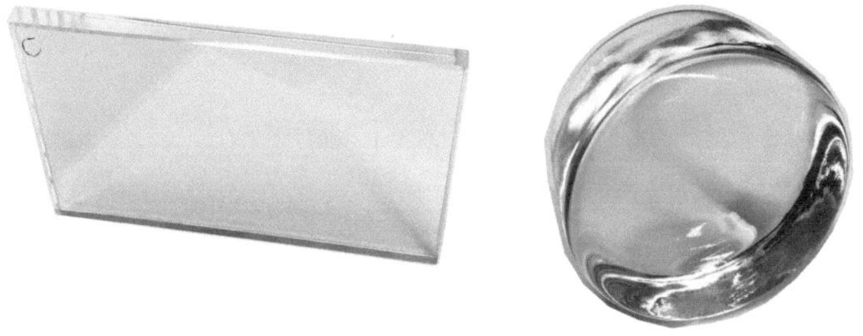

(a) (b)

Figure 1.3: Different types of test objects: (a) a glass plate with a simple geometry is a cooperative test object; (b) a glass gob whose surface structure shows many variations is an uncooperative test object.

2. **Accuracy:** The system's ability to reveal material defects, i.e., the maximum contrast-to-noise ratio (see Section 6.1.3) achieved by the system (the higher, the better).

3. **Speed:** The number of test objects that can be inspected in a certain amount of time (the more, the better).

4. **Adaptability:** The ease of adapting the inspection system for new types of test objects, e.g., with a different geometry (the greater, the better).

5. **Robustness:** The degree to which the properties of the test objects (e.g., their shape) and those of the inspection system (e.g., accuracy of the positioning of the test object) may vary without notably influencing the performance of the inspection system (the greater, the better).

These criteria will be used throughout the thesis whenever a visual inspection system or method has to be evaluated.

1.1.3 An example inspection system

As an example, this section introduces and explains the *Purity* system, that was developed by Fraunhofer-IOSB [Har08; Har09; Mel10; Mey15]. This system represents the state of the art (cf. Chapter 2) with respect to the detection of absorbing and scattering defects in test objects with a complex shape. The Purity system is also evaluated with respect to the previously introduced criteria at the end of this section and serves as a benchmark for the novel methods introduced in this thesis.

Figure 1.4 illustrates the optical setup of this inspection system. The test object is illuminated by different light sources and is observed by a telecentric camera which allows measuring the sizes of found defects. Either an image series is obtained in a time-sequential manner, where each illumination source is turned on separately, or color multiplexing is applied by using light sources emitting different spectra and a color camera. A collimated bright field illumination (blue rays) is directed parallel to the optical axis of the camera by a beam splitter. When being transmitted through the test object, the light rays are refracted and their direction of propagation is changed. The retroreflector reflects the incident light rays back along their original optical path. By this means, the refraction effects of the test object are mitigated what makes the system applicable to objects with a complex geometry. After passing the beam splitter, the light rays reach the camera and yield the transparency profile of the test object. Any absorbing defect (yellow dot) will result in a locally reduced intensity in the acquired image. Scattering defects are not revealed by the bright field illumination since their optical effects are also inverted by the retroreflector. Therefore, various dark field light sources are arranged around the test object (green light rays). The concept of dark field illumination states that, in the

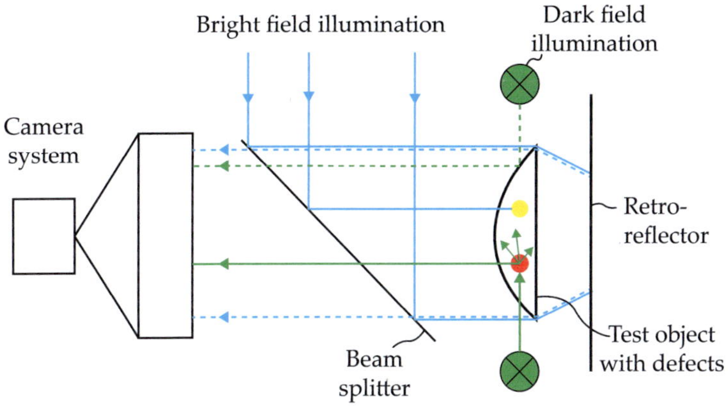

Figure 1.4: Optical setup of the Purity inspection system.

case of a defect-free test object, no light reaches the camera. If the test object is affected by a scattering defect, the incident light will be scattered into multiple directions (solid green rays) what results in high sensor signals revealing the defect. However, depending on the test object, the arrangement of the dark field light sources has to be specifically designed and adapted to the objects of interest to avoid direct reflections of the light sources into the camera (dotted green ray). Such reflections result in blind spots of the image acquisition system. For the inspection of flat transparent objects, such reflections can easily be avoided. For complex-shaped objects it is a tedious and time-consuming process to adapt the system to a new type of test object.

The Purity inspection system can be discussed by means of the criteria introduced in the previous section:

1. **Versatility:** Due to the different illumination channels, both absorbing and scattering defects can be revealed.

2. **Accuracy:** For regions of the test object that are free from reflections, high accuracies can be reached (see Section 6.3.4). In the other regions, the disturbing reflections greatly reduce the accuracy, so that the detection of defects is impaired.

3. **Speed:** For the time-sequential acquisition of an image series, the inspection time depends on the required exposure time and the number of channels, i.e., of different light sources. For a configuration with one bright field illumination and 20 dark field light sources, the acquisition time will be in the range of 2 s.

4. **Adaptability:** For a new type of test object, an expert has to adapt the arrangement of the dark field light sources by means of empirical experiments.

5. **Robustness:** Depending on the type of test object, small variations of the test object might be tolerable. Since the positions of the dark field light sources depends on the position of the test object, the system will fail for misalignments of the test object.

1.2 Core idea: Light fields for the visual inspection of transparent objects

In the domain of computer graphics and computer vision, the so-called *plenoptic function* is a frequently used tool to conceptually describe the light transport of a scene. Adelson and Bergen [Ade91] initially defined the 7-dimensional plenoptic function

$$\rho(x, y, z, \theta, \varphi, \lambda, \tau) \tag{1.5}$$

as the radiance, i.e., the intensity of light rays with a wavelength of λ incident at a point $(x, y, z)^\mathsf{T}$ with an angle of incidence of (θ, φ) at time τ. It has to be noted that the plenoptic function assumes geometric optics, i.e., that all relevant geometrical structures are larger than the light's wavelength and that there are no diffraction effects. Due to the high dimensionality of the domain of ρ, it is usually infeasible to capture the complete plenoptic function.

However, there are cases with practical significance in which a merely four-dimensional section of the plenoptic function is sufficient for many applications [Lev96]. If the scene of interest is free from any occlusions between the objects of interest and the observer, the radiance transported along a ray starting from point $(x, y, z)^\mathsf{T}$ on an object's surface in direction (θ, φ) does not change, which is why one of the three spatial parameters is redundant. Furthermore, if the scene is static, i.e., constant with respect to time, and if the light's wavelength is neglected, the two parameters λ, τ are of no importance. The resulting four-dimensional function is called a *light field*

$$L(x, y, \theta, \varphi) = \int \rho(x, y, z, \theta, \varphi, \lambda, \tau) \mathrm{d}\lambda|_{z=\mathrm{const.}, \tau=\mathrm{const.}} , \tag{1.6}$$

with respect to a certain plane $z = \mathrm{const}$. Another common parametrization of light fields is to describe the light rays by their intersections with two parallel planes at a distance larger than zero. However, in this parametrization, rays propagating parallel to the two planes cannot be represented. If not explicitly stated differently, two spatial coordinates and two angular coordinates are used to parametrize the light fields in the remaining parts of this thesis.

Many of the relevant material defects contained in the DIN ISO 10110 standard (see p. 2), e.g., enclosed bubbles, striae, scratches, etc., have a strong effect on the distribution of the propagation direction of the transmitted light rays. These defects scatter incident light rays into multiple directions or abruptly redirect them away from their original propagation direction [Boh07; Hul57]. But, due to refraction effects, also the shape of transparent objects manipulates the light's propagation direction. Figure 1.5 illustrates these considerations via light fields showing the distribution of the deflection angles of parallel light beams after being transmitted through a transparent plano-convex lens that is partially affected by material defects. In defect-free regions of the test object, the light fields show a concentrated peak that is shifted only slightly with respect to adjacent rays due to the smooth surface gradient (rays 1, 2, 4 and 6). In contrast, scattering defects scatter the incident light into multiple directions, resulting in broad intensity distributions (ray 3). Material defects like cracks, large scratches, or incomplete shapes can lead to local abrupt changes of the propagation direction (ray 5). These considerations suggest that light fields contain much information relevant for finding material defects in transparent objects. The way humans manually inspect transparent parts also supports these findings. Instead of just observing a transparent test objects in front of a uniform bright light source, humans tend to rotate the object and look at it from different angles. By this means, they capture a light field in a time-sequential manner. The observer will judge abrupt changes between the different viewpoints as an anomaly that could constitute a material defect. Consequently, the core idea followed by this thesis is to adequately employ the concept of four-dimensional light fields in all appropriate components of a machine vision system for the visual inspection of transparent objects.

1.3 Scope and objectives

The main goal of this thesis is to introduce approaches and methods suitable for developing novel visual inspection systems or to improve existing ones. The target application is the visual inspection of transparent objects, which are uncooperative in the sense described in Section 1.1.1, with respect to absorbing and scattering impurities. In this scenario, the information about the presence of such material defects is contained in the direction of propagation of the light, i.e., in the light field, as pointed out in the previous section. Conventional machine vision systems not considering the light's direction of propagation fail at this task with respect to the criteria specified in Section 1.1.2. Hence, the contributions of this thesis introduce the light field concept into all main components of a visual inspection system, the illumination, the image or signal acquisition device, and the image processing unit.

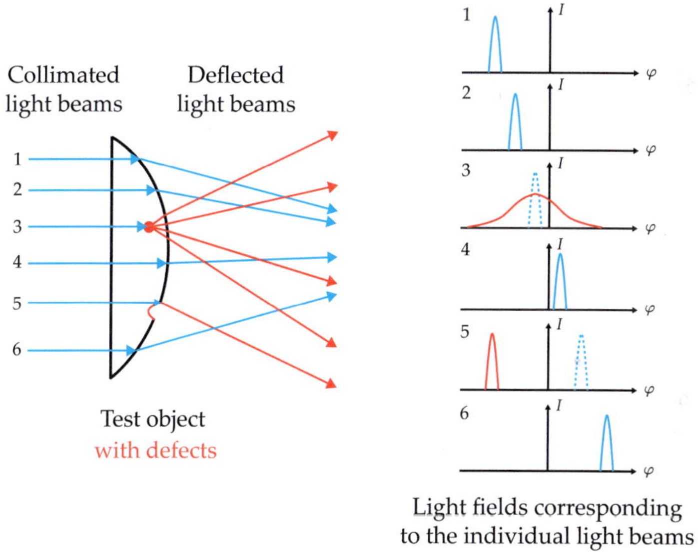

Figure 1.5: Manifestation of material defects in light fields: collimated light beams illuminate a plano-convex lens with an enclosed air bubble (ray 3) and a surface defect (ray 5). Both the object's surface gradient and the material defects affect the distribution of the transmitted light's direction of propagation. The light fields on the right-hand side show the vertical component of the intensity distributions of the individual light rays after being transmitted through the test object. In the cases of ray 3 and ray 5, the dashed lines indicate the peak location for the defect-free case at the corresponding positions.

The proposed novel methods are developed in accordance with the criteria listed in Section 1.1.2 as high-level guidelines. The visualization of material defects while simultaneously suppressing all the nominal features and structures of the test object constitutes the core of this thesis. A successive classification of the visualized defects with respect to different defect categories (e.g., bubbles, scratches, etc.) is not pursued. However, appropriate techniques could be appended to the processing pipeline in future research (see Chapter 8).

1.4 Scientific contributions

The main contribution of this thesis is constituted by the concepts of acquiring and adequately processing light fields for the visual inspection of transparent objects. By adapting sophisticated distance measures for comparing the light deflection distributions of adjacent object points, even test objects with complex shapes can be inspected without having to tediously design a specialized machine vision system. Novel and efficient algorithm designs result in short computation times, so that the achievable inspection speed is fast enough for the requirements of industrial inspection applications.

This thesis contributes to the state of the art with respect to all the components of a visual inspection system for transparent objects and materials. In detail, the following novelties are introduced:

- **Light field illumination** [Mey18a]
  A light field generator is optically realized by combining a microlens array with a high-resolution two-dimensional display (see Section 3). By this means, some of the display's spatial resolution is sacrificed to gain angular resolution, i.e., to be able to control the angle of the emitted light beams. Furthermore, a calibration procedure is developed that yields the

correspondence between the controllable pixels of the underlying display and the spatial location and emission angle of the corresponding light beams. A prototype is built that validates the concept.

- **Light field acquisition**

  A novel optical design of a light field camera is introduced [Mey16d]. By realizing a telescope-like optical setup with two lenses sharing a common focal plane, the deflection angle of captured light bundles with respect to the optical axis can be measured with a constant resolution over the whole field of measurement (see Section 4.2.2). This is not possible with conventional light field cameras. The $4f$-light field camera is realized as a plugin for the simulation framework Mitsuba in order to simulate physically realistic light field images.

  A transmission-based laser scanner capable of acquiring high-resolution light fields of transparent test objects is presented [Mey16a]. Due to the sensor's high acquisition speed, its scalability, and the fact that it can be installed in-line in a production system, it is well suited for industrial inspection tasks (see Section 4.4.1). Via an additional position sensitive detector that can optionally be integrated into the sensor, even more precise measurements of the light's deflection angles can be obtained and the incident light field can be captured faster and in a more compact representation if necessary. The laser deflection scanner is realized as a prototype and successfully evaluated by means of several experiments.

  An optical setup based on the combination of a camera and a two-dimensional programmable light source is proposed for capturing the light transport matrices of transparent objects [Mey16c; Mey16e; Mey17c]. Furthermore, two features are introduced which are calculated from the acquired light transport matrices (see Section 4.5 and 6.4). These features are designed to highlight any scattering material defects in transparent objects. Several simulated experiments validate the proposed approach.

- **Machine vision simulation framework** [Mey16b]
  Several plugins for the physically based raytracer Mitsuba are introduced (see Chapter 5). Based on these plugins and the Mitsuba renderer, virtual machine vision systems can be designed and the digital images of the employed sensors can be computationally simulated. By this means, the design of machine vision setups is streamlined and it is possible to evaluate novel system concepts in a rapid prototyping manner.

- **Light field processing methods for transparent object inspection**
  A framework for processing light fields for defect visualization is presented [Mey16d]. For this purpose, the gradient operator known for detecting edges in gray-value images is extended to deflection maps, i.e., light fields of transparent objects acquired under special illumination (see Section 6.1). By employing suitable distance functions, the nominal features of the investigated object can be suppressed while material defects are extracted with high contrast.

  An adaptation of the Generalized Cramér-von Mises Distance [Mey17b] is introduced and successfully employed as such a distance function (see Section 6.1.2). For the efficient calculation of this distance, a fast algorithm is developed based on the concept of summed area tables [Mey17a]. By this means, the necessary computation is reduced from $\mathcal{O}(n^{2.5})$ to $O(n^{1.5})$ for an input problem of size n.

An approach based on vector analysis is presented, that allows to process the signals of a position sensitive detector installed in a laser deflection scanner to reveal larger material defects (see Section 6.2). By this means, a fast inspection requiring only 1.26 s per test object can be realized.

Two image fusion algorithms are proposed [Mey18b] that take different amounts of automatically learned information about a visual inspection task into account (see Section 6.3). Depending on whether only knowledge about the object's nominal geometry and structures is available or additional information about the appearance of material defects can be acquired, the algorithms optimize a specifically designed signal-to-noise ratio. During the image fusion process, this optimization leads to suppressing the intended object features while highlighting material defects, visualizing them with high contrast.

- **Inverse light field illumination** [Mey16f]
  An optical setup is proposed which illuminates a transparent test object with a programmable light field using the introduced light field generator and observes the other side of the object with a telecentric camera (see Chapter 7). With an intact defect-free test object, a learning procedure based on a coded illumination series allows determining an inverse reference light field. When illuminated with that inverse light field, a defect-free test object transforms the light field into light bundles oriented parallel to the optical axis that reach the sensor of the telecentric camera and yield a bright image. Any material defects that absorb light, or scatter or redirect it in another direction, will lead to rays that are not propagating parallel to the optical axis. These rays will be blocked by the camera's telecentric stop, which will result in dark image regions. This approach

allows inspecting the whole volume of even complex-shaped test objects by acquiring only a single image.

All ingredients of the contributions, i.e., optical setups, processing methods, and other artifacts, are rigorously evaluated either by simulated experiments or—if applicable—also by conducting real experiments.

Together, the contributions constitute a toolbox of light field methods for the development of novel visual inspection systems for uncooperative test objects. Depending on the application, the toolbox components can be combined in various ways to solve the inspection task: The methods for processing deflection maps could be employed together with the $4f$-light field camera for the inspection of small test objects with diameters in the order of a few millimeters up to 15 cm (e.g., gobs, lenses), or with the laser deflection scanner for the inspection of larger test objects with diameters of 50 cm and greater (e.g., headlamp covers of automobiles or washing machine door glasses). The described Purity system (see Section 1.1.3), has already been extended with the image fusion algorithms to achieve a higher accuracy and to improve the system's robustness. The system has been successfully installed in an industrial production line for complex-shaped plastic lenses. A novel visual inspection system based on the inverse light field illumination method could be realized by using the developed light field generator. With this system, small test objects with dimensions in the order of 50 mm $\times$ 50 mm, e.g., small glass bottles, could be inspected.

1.5 Thesis structure

Up to now, Chapter 1 has introduced the basics of an automated visual inspection system and its design, explained the important difference between cooperative and uncooperative test objects, stated the core idea of the thesis, viz., using the concept of light fields for the visual inspection of transparent objects, and has summarized the scientific contributions of this thesis.

Chapter 2 will give a thorough overview of the related state of the art. Since this thesis contributes to several fields which are all connected to the visual inspection of transparent objects, Chapter 2 is deeply structured to systematically group the related work.

In Chapter 3, the optical realization and calibration of a light field generator is described.

Chapter 4 covers all optical setups and concepts suitable for acquiring light fields in the context of an industrial inspection. Furthermore, the idea of light deflection maps is introduced, which is a key component for the later processing approaches.

Chapter 5 introduces a framework that allows to simulate the images of the sensors of a simulated machine vision system. Besides a description of the employed Mitsuba renderer, the appended framework components are covered.

Chapter 6 contains all presented methods and algorithms for processing light fields captured from transparent objects in order to extract any material defects. Suitable methods for processing deflection maps are developed. In addition, image fusion strategies based on the optimization of a specifically designed signal-to-noise ratio are described. Two features are introduced, which can be extracted from light transport matrices, and which especially highlight scattering material defects.

Chapter 7 covers the approach of inspecting transparent objects by inverse light field illumination. After explaining the method and the underlying optical setup, both simulated and real experiments are described and evaluated.

Lastly, Chapter 8 closes the thesis with a summary and a conclusive evaluation of the contributions and by providing an outlook for future research questions.

2 The state of the art

This thesis contributes to multiple components of a machine vision system: the illumination part, the image acquisition device and the image processing methods. Therefore, the related state of the art also has a wide variety originating from the work of many researchers and developers from different fields. In order to provide a well-structured overview of the important work related to the content of this thesis, this chapter discusses separately the current state of the science in Section 2.1 and the state of the technology in Section 2.2. Section 2.1 covers all major relevant research reports by other scientists and Section 2.2 provides an overview of the related existing products and machine vision components. Both sections are organized to match the high-level grouping of the scientific contributions of this thesis.

2.1 The state of the science

Many research groups are involved in the topics covered by this thesis. The following sections will briefly summarize their contributions and discuss how they are related to this thesis.

2.1.1 Transparent object inspection

The inspection or reconstruction of transparent test objects is an actively studied field. As shown in Section 2.1.1.1, many publications in this area treat the automated visual inspection of flat transparent panes. Other types of objects of great interest are the glass or plastic bottles used as containers for liquids. The corresponding literature is discussed in Section 2.1.1.2. Only a few related publications can be found dealing with the visual inspection of uncooperative transparent objects (see Section 2.1.1.3) and their reconstruction (see Section 2.1.1.4).

2.1.1.1 Flat transparent objects

The production of flat glass constitutes a major sector of the glass industry, worldwide. Flat glass panes are employed to build windows for houses or even completely transparent walls for modern houses or offices. Since their main purpose is to provide a clear and undistorted view through the windows or walls, these products have comparatively low quality requirements: they only have to be free from contaminants that are visible to the naked eye, e.g., large light-absorbing or scattering particles, severe scratches, and disturbing color casts. The index of refraction may vary to even a large extent, since humans are not sensitive to the resulting phase effects. Also, flat panes made of transparent plastics are widely used in different kinds of products, e.g., as covers for smartphones or other display devices.

Most of the related literature presents inspection systems relying on the combination of bright field and dark field illumination to visualize and detect both absorbing and scattering material defects [Ada09a; Ada09b; Ada10; Cha16; Lia16; Yu11]. Since in these publications simple thresholding operations are reported to be sufficient for the detection of defects, other authors have presented more complex methods, such as principal component analysis [Li14], a wavelet transform with specially designed wavelets [Akd15], or a convolution with Gabor filter kernels [Jeo11]. In [Li15], moiré patterns have been transmitted through the glass plates, observed, and appropriately processed to successfully identify material defects.

2.1.1.2 Bottles and food containers

Bottles for liquids and other food containers made out of transparent materials also represent a class of test objects whose visual inspection is often studied. Although bottles are uncooperative test objects, due to their complex geometry, classical inspection approaches such as bright field and dark field setups can be successfully applied. This is because the relevant defects are rather macroscopic, e.g., broken edges, large cracks, etc.

In [Wan00], wineglasses are inspected by observing the fringe structure of a transmitted moiré pattern and by employing Gabor filters to find irregularities. Inspection systems for glass bottles are presented in [Dua04a; Dua04b; Dua07; Tox13]. There, every test object is imaged under different illumination modalities in order to obtain suitable inspection images of the bottle's mouth, its wall, and its bottom. Each of these images is then preprocessed to mitigate the effects of small variations in placement, and then specially tailored image processing pipelines are employed to detect any material defects.

2.1.1.3 Uncooperative objects

As pointed out in the introduction (see Section 1.1.1), the visual inspection of uncooperative transparent objects is a challenging task. Little related research has been performed on this topic.

Martínez et al. present a so-called *active binary lighting* system that allows inspecting even complex objects, e.g., the plastic covers of automotive headlamps [Sat09; Sat12]. To this end, they employ a robotic platform moving a camera system around the test object.

Their illumination approach consists of a monitor displaying a moving binary black and white stripe pattern. They require the test object to have no cavities and that the front and back surfaces are approximately parallel to each other. Any scattering defect on or inside the test object will produce a deflection of the rays. If the monitor shows the edge between a white and a black stripe directly behind the defect, the deflection causes some of the light of the adjacent bright stripe to produce a bright spot in the camera image. An absorbing defect will result in a dark structure in the image if a bright stripe is shown behind the defect. A series of images is acquired while phase shifting the binary pattern. After computing, for each pixel, the average over the series of images of its gray values, pixels showing defect-free regions have a gray value corresponding to the mean of equally many bright and black stripes, whereas the gray values of pixels belonging to material defects are either darker (absorbing defects) or brighter (scattering defects). By capturing multiple image series while moving the camera around the test object, the whole test object is successively scanned. Because of the robot platform, the setup can inspect even objects with a complex geometry. However, this also leads to a long inspection time and impairs the adaptability of the method since a sensor planning strategy has to be found for every type of test object.

2.1.1.4 Reconstruction of transparent objects

The reconstruction of the three-dimensional shape of a transparent object by means of computer vision approaches has been studied by several researchers. Since usually the work is not focused on the visual inspection of the reconstructed objects, only a few representative publications are described in the following. A comprehensive review is provided in [Ihr10].

In [Wet11b], Wetzstein et al. present a specially designed illumination device consisting of a background illuminated two-dimensional color pattern in front of which a microlens array is placed at a distance of the focal length of the single lenses. Hence, the visible parts of the color pattern depend on the direction of observation. By placing a printed image of a color wheel under each microlens, e.g., of the hue-saturation-value color space, the two-dimensional emission angles of the light rays is encoded. An additional intensity gradient over the complete pattern allows encoding the first spatial coordinate of the whole illumination device. The second spatial coordinate can be computationally recovered through geometric constraints. In order to perform the reconstruction of a transparent object, it is illuminated with this four-dimensional light source and the transmitted light is captured with a calibrated color camera with a common entocentric lens. By assuming a pinhole camera model, every camera pixel is illuminated by light rays with a certain direction. The captured color information is decoded into the point of origin and emission angle of the corresponding light ray emitted by the light source. Based on this information, the authors succeed in reconstructing the surface shape of a thin transparent object.

Ma et al. illuminate transparent objects or other refractive phenomena of interest (e.g., gas flows, liquids, etc.) with a collimated light beam in which the spatial location of the contributing ray bundles can be encoded via multiple intensity patterns [Ma14]. This illumination is mainly realized by means of a beamer based on a digital micromirror device. The transmitted light is split into two optical paths and is observed by two cameras at different effective distances to the object. By determining the spatial coordinates of the illuminating ray bundles for each camera pixel in both camera images and by calculating the displacement between the coordinates of the pixels from the two cameras illuminated by the same ray bundle, information about the ray bundle's direction of propagation can be obtained. Since the system is calibrated, the propagation directions together with the spatial positions of all ray bundles allow reconstructing the surface of a transparent object or the refractive index distribution of a gas flow or liquid.

2.1.2 Light field illumination

In [Tak16], Takeuchi et al. introduced a programmable ambient light field illumination source called *AnyLight*, which relies on a spatially programmable light source in combination with an array of lenses. AnyLight consists of an LCD panel that is positioned in front of an array of high power LEDs employed for background illumination. A customized array of lenses is mounted on top of the LCD panel. Opaque structural elements are positioned between the individual lenses in order to reduce cross talk. The reported prototype has a comparably low spatial resolution of 3.58 $\mathrm{mm} \times$ 3.58 mm. However, since the authors target the application of illuminating architectural environments, this spatial resolution is sufficient. The reported experiments show that the approach is suitable for adequately lighting indoor scenes.

In [Hua14], Huang et al. realize a light field display by mounting a pinhole array on top of a high resolution smartphone display. The pinhole array acts as a parallax barrier. If a pixel is turned on, nearly all of its light is blocked by the pinhole mask, except for a small portion, which propagates through the corresponding pinhole. By this means, both the spatial location of the light bundle's origin and its direction of propagation can be controlled. The authors successfully show that their light field generator can be used to emit visual content that is optically preprocessed, so that observers affected by visual aberrations can see a sharp image of the content without having to wear any eyeglasses. A drawback of their optical setup is the reduced light efficiency caused by the parallax barriers.

Wetzstein et al. [Wet11a] employ a high resolution two-dimensional display in concert with multiple LCD layers serving as transmission masks in order to realize a light field display. By adequately setting the transmittance pattern of the individual LCD panels, they are able to emit light fields with a high spatial and angular resolution. The ensemble of transmittance patterns acts like a filter, letting pass only those light rays that have the desired point of origin and direction of propagation. The more layers are used, the more accurately a desired light field can be emitted, but also the more light is lost during transmission. If only a few layers are used, there will be more unintended light rays present in the generated light field, leading to artifacts with respect to the intended light field. Similar approaches are pursued by [Lan11; Wet12], where an even more accurate perception of the desired light field is achieved by means of a temporal integration at the observer. It should be noted that these light field generators are targeted mainly to displaying three-dimensional content to human observers, and not to machine vision applications.

2.1.3 Light field acquisition

The idea of light field imaging was first described by Lippmann in 1908. In [Lip08], he already employed the principle of using a microlens array in front of a photosensitive film to conserve the directional information of the captured light bundles. This concept was adopted by Ng to develop the first hand-held light field camera in 2005 [Ng05], by mounting a microlens array in front of the sensor of a commercially available digital single-lens reflex (DSLR) camera.

Following this approach, several light field cameras have been produced. Since these cameras are related to the state of the technology, they are described in Section 2.2.3.

Liang et al. create a light field camera by introducing a programmable aperture into a DSLR camera [Lia07; Lia08]. They realize a programmable aperture, e.g., by attaching an LCD display to the image-sided end of the camera's lens system. This LCD display acts as a programmable transmission mask whose patterns apply weights to the contribution of the captured light bundles to the image formed on the sensor. By acquiring multiple images of a static scene for different binary mask configurations, a series of multiplexed light fields is obtained. A successive computational demultiplexing eventually yields the sought light field. Since no microlens array is used in this approach, the recorded light fields have a high spatial resolution but require the observed scene to be static for a certain period of time, as multiple images have to be acquired.

In [Ant16], Antipa et al. present a light field imaging approach based on a thin light diffuser plate that is mounted in front of a camera sensor. Incident light rays are scattered by the diffuser plate and reach several spatially distributed sensor pixels. By calibrating the system and by adequately modeling the light transport, they succeed in providing an algorithm capable of reconstructing, based on a single acquired image, an approximation of the incident four-dimensional light field. In the reported experiments, the reconstructed light field images have a spatial resolution of 170×170 and an angular resolution of 11×11.

Camera arrays are another frequently employed approach for acquiring light field images [Yan02]. Several common cameras are arranged in a two-dimensional grid and capture images of the scene of interest. Due to this arrangement, every camera observes the scene from a different perspective, which allows combining the single camera images into a light field image. Such acquisition systems have a high spatial resolution at the cost of large geometric system dimensions. By additionally making use of a pre-trained light field dictionary, the angular resolution of camera array-based systems can be further increased [Cao14].

Another approach for capturing light fields for the visual inspection of transparent objects is the so-called schlieren deflectometer [Sud13; Sud15] employed by Sudhakar et al. This optical setup consists of a pivoting collimated light source illuminating the test object from one side together with a telecentric camera observing the object from the other side. By capturing a series of images from multiple collimated light beams, tilted by different but known angles, a light field image of the test object can be successively acquired. By additionally employing the concept of compressive sensing, the amount of data needed to represent the usually sparse light field of a transparent object can be reduced. The authors target contact lenses as test objects and further process the acquired light fields to determine the spatially resolved optical power of the lenses.

2.1.4 Simulation of machine vision setups

In the domain of computer graphics, the simulation of the image formation process of cameras is an intensely studied field of research. Different rendering frameworks have been introduced, which are focused on implementing an image formation as close to real world physics as possible, by taking all relevant optical effects into account [Pha16]. Since most of these frameworks are targeted to artistic applications, they are not directly suitable for simulating machine vision systems. However, different researchers have successfully employed them to streamline the design of machine vision setups.

In [Nür16], Nürnberg et al. employed a rendering framework in order to optimize the parameters of a computational camera used for depth estimations.

Mohammadikaji et al. [Ber16; Moh16; Moh17] presented an uncertainty propagation framework for finding the optimal arrangements of a laser triangulation setup by means of precisely simulating the laser's speckle effects and the reflectance properties of the investigated test object.

Irgenfried et al. [Irg11; Irg14] developed a software framework capable of optimizing the parameters of image processing routines involved in a visual inspection process. Therefore they simulated the inspection scene using both realtime and photorealistic renderers. They also compared their simulation results with real measurements, and reported that by adequately modeling a machine vision setup, the simulation results are close to the real camera images [Irg17].

2.1.5 Light field processing

The idea of describing light transport by means of light fields and to acquire and process them can mainly be found in the computer vision community. In this area, there are many research reports showing how light fields can be employed to gain additional information about an observed scene or how to computationally generate artificial views of it. In his dissertation, Ng covers all the required mathematical background and describes several examples of applications of light field processing [Ng06]. One of his main contributions is an algorithm that allows obtaining images for different planes of focus based on a single light field image. By performing most of the required calculations in the Fourier domain, the algorithm's computation time is notably reduced compared to other approaches.

Light field images also allow estimating the depths of the observed objects, i.e., their distances from the camera [Tos14]. Shao et al. present a segment-based depth estimation method based on graph cuts [Sha15]. In [Wan14], Wanner and Goldlücke introduce variational methods for depth estimation and for obtaining super resolution images.

In [Yao16], Yao et al. introduce a method that allows computationally changing the point of view of an observed scene. By employing techniques of Ng's Fourier-based algorithms, they can achieve real-time performance.

2.1.5.1 Light field processing for visual inspection

Although the processing of light fields has its main applications in the domain of computer vision and computer graphics, there are also some machine vision applications that have been reported.

Svorad Štolc et al. capture the light fields of printed circuit boards (PCBs) transported on a conveyor belt in order to inspect them for completeness [Što14]. For this purpose, they employ multiple line scan cameras, each tilted by a certain angle so that they observe a common line on the conveyor belt from different viewing angles. By this means, they are able to reconstruct the light fields reflected by the objects and use them to calculate all-in-focus images or depth maps. They showed that the resulting data could successfully be employed for the inspection task.

In [Sou15], Soukup et al. use a light field camera in concert with a variable direction dome illumination in order to partially acquire the bidirectional reflectance distribution function of so-called diffractive optical variable image devices (OVID). Such OVIDs are frequently used as anti-counterfeiting measures on bank notes, since their reflectance is highly dependent on the angle of incidence of the illumination and on the angle of observation. Because of the variable illumination direction and the angular resolution of the sample's reflection achieved by the light field camera, they were able to define adequate features suitable for discriminating genuine bank notes from counterfeit ones.

2.1.6 Light transport matrix acquisition and processing

As thoroughly described in Section 6.4, light transport matrices provide the possibility of completely describing the light transport of a static scene. A light transport matrix contains the contribution of every light source present in the scene to every camera or sensor pixel observing the scene. Different setups and methods have been formulated to acquire light transport matrices and to further process them for different purposes. The main common challenge of these methods is to obtain a compressed representation of the light transport matrix, as it would require infeasible amounts of computer memory for state-of-the-art camera systems.

The light transport matrix of a scene enables synthesizing images of the scene under different lighting conditions, i.e., for different intensities of the present light sources. In [OTo10], an optical setup consisting of a camera and a video projector and an illumination and processing strategy is employed that allows acquiring a scene's light transport matrix. In this approach, multiplications with the light transport matrix are performed in the optical domain by projecting images, i.e., vectors, onto the scene and by simultaneously capturing images with the camera, which correspond to the result of the vector-matrix-multiplication of the projected image with the unknown light transport matrix. By coupling this approach with so-called Kernel Nyström-methods [Wan09], it is possible to obtain an approximation of the light transport matrix based on its eigenvectors corresponding to the largest eigenvalues.

In [OTo14], O'Toole et al. also employ a camera–projector setup which they calibrate similar to a stereo camera system, in order to obtain epipolar relations between the two devices. By means of this setup, and the appropriate acquisition and image processing algorithms, they can acquire images of even non-static scenes that are formed either only by direct or indirect light transport. Light rays directly reflected into the camera by the scene constitute the direct light transport and light rays that undergo two or more reflections before being captured by the camera constitute the indirect light transport. Especially those images consisting only of indirect light transport reveal interesting sights onto the scene, e.g., the veins (blood vessels) under the skin of a human arm are visualized more clearly due to the emphasizing of subsurface scattering effects.

Light transport matrices acquired with a camera–projector setup also allow enhancing the perceived quality of projected images or videos. Miyagawa and Kinebuchi show that it is possible to mitigate radiometric distortions affecting the projected images caused, e.g., by a screen that is not colored completely white [Miy17]. By first detecting such distortions based on an analysis of the acquired light transport, the projected images can be inversely transformed so that they appear as intended when projected onto the screen.

2.1.7 Inverse illumination

The idea of using a specially designed illumination that inverts certain optical effects has been applied in different variations in the field of automated visual inspection.

In [Bot03; Pös12] fringes are projected onto the test object, which are inversely adapted to the test object's geometry. The inverse pattern is obtained by means of a defect-free sample of the test object. If the determined pattern is projected onto a test object whose 3D-structure is free from shape defects, it transforms the incident distorted fringe pattern into an easily recognizable regular pattern of parallel fringes. Conversely, any material defect affecting the object's geometry will result in a distorted pattern, indicating a deviation from the nominal state. This approach has also been used for the deflectometric inspection of specular surfaces [Wer07].

There are also inverse illumination techniques in the field of comparative digital holography [Bau06; Ost02]. The coherent optical wave field reflected by the reference object is acquired by means of digital holography. Illuminating a test object with this coherent wave field directly visualizes differences between the two objects' shapes.

In [Gru10], Gruna et al. present another inverse illumination approach that is based on light transport inversion in a camera–projector system. Their system is able to visualize both the differences between the shapes and differences between the reflectances of a reference object and a test object, with high contrast in the inspection images.

2.2 The state of the technology

Regarding some of the topics covered by this thesis, there are also fully developed technologies commercially available as products. This section reviews the state of the technology that is currently employed or provided by the industry. The following content is organized in accordance with the previously described state of the science and the overall structure of the thesis. As far as the present author's knowledge extends, there is no usable mature technological solution or product for light field processing and inverse light field illumination.

2.2.1 Transparent object inspection

As mentioned in the introduction to Chapter 1, the inspection of transparent materials is an important topic for many industrial fields. Hence, various companies have developed machine vision systems capable of performing the different required inspection tasks.

2.2.1.1 Flat transparent objects

Visual inspection systems for flat glass are provided, e.g., by the companies *Viprotron* [Vip], *ISRA (glass vision)* [ISR], *Softsolution (Glass-IQ)* [Sof] and *Dark Field Technologies* [Dar]. Most of these technologies rely on straightforward optical setups based on a combined arrangement of a bright field and dark field illumination. The solution provided by Dark Field Technologies additionally uses a laser scanner for high intensity dark field illumination. An inspection speed of up to 80 $\mathrm{m/min}$ for glass plates with a maximum width of 3 m can be realized (source: Viprotron).

The Purity system already described in the introduction (see Section 1.1.3), that was developed by Fraunhofer-IOSB [Har08; Har09; Mel10], makes use of three illumination sources. The system reveals absorbing contaminants, scattering impurities and also surface defects in flat transparent objects.

2.2.1.2 Glass and plastic bottles

Transparent glass or plastic bottles can be automatically inspected, e.g., by the systems developed by *Optische Prüfsysteme Dr. Günther* [Gün], *Antares Vision (GIS)* [Ant], *miho Inspektionssysteme (David 2)* [mih] and *IRIS inspection machines* [IRI]. The fastest inspection speed is claimed by the David 2 system (miho Inspektionssysteme) with 72,000 test objects per hour.

2.2.1.3 Glass tubes

For the production of optical fibers or fluorescent lamps, glass tubes are an important intermediate product that has to meet high quality requirements. Adequate visual inspection systems are offered by *Opos* [OPO] and by *FiberVision* [Fib]. Opos employs laser scanners with a bright field and a dark field channel to detect both absorbing and scattering contaminants. The system also allows obtaining dimensional measurements of the test object. The *FiberVision* systems are based on multiple cameras combined with a bright field illumination, enabling an inspection for absorbing structures or material cracks.

2.2.1.4 Uncooperative objects

The Purity system mentioned before (see Section 1.1.3) is also capable of inspecting test objects with complex shapes, like glass lenses for optical systems or plastic lenses for automotive headlights. However, depending on the test object, it requires a multichannel illumination component that is specifically designed and adapted to the objects of interest. Hence, it is a tedious and time-consuming process to adapt the system to a new type of test object.

In their *NIMO* and *PMTF* product series, the company *Lambda-X* realizes the principle of the schlieren deflectometer already mentioned in Section 2.1.3 and further explained in Section 4.3.2 for the visual inspection of contact lenses [Lam]. The system allows obtaining spatially resolved measurements of the optical power of inspected contact lenses.

Industrial Vision Systems [Ind] and *Shelton Vision* [She] also offer inspection systems for contact lenses. Their solutions are based on cameras that acquire image series of the test object under variable illumination and variable distances of the plane of focus (see the presentation of the thin lens camera on p. 61). An image fusion procedure processes the image series and yields sharp images over the whole of the volume of the test object, which are free from disturbing reflections and allow a reliable detection of defects.

2.2.2 Light field illumination

At the time of the creation of this thesis, there are no commercially available technologies allowing the emission of user definable, fully controllable, four-dimensional light fields. However, there are some devices that enable users to view three-dimensional content without having to wear special glasses, e.g., *Nintendo 3DS* [Nin], *VisuMotion* [Vis], *SeeReal Technologies* [See] and *REALEYES* [REA]. However, these products mainly rely on providing two separate views of a virtual scene for the two eyes of a human observer in order to create a 3D-impression. They do not emit a fully customizable four-dimensional light field.

2.2.3 Light field acquisition

Two major companies are known to produce or to have produced light field cameras. *Lytro* manufactured the famous hand-held light field camera *Illum* for the consumer market [Lyt][1]. However, the company shut down operations in 2018.

The German company *Raytrix* [Ray] develops light field cameras targeted for industrial applications, especially for the inspection of printed circuit boards (PCBs) by means of the depth information that can be extracted from acquired light field images.

[1] The website of Lytro is no longer accessible.

2.2.4 Simulation frameworks

Currently, there are no (commercially) available simulation tools that allow simulating a complete machine vision setup for visual inspection applications. However, there are software tools, e.g., *Zemax* [Zem], that can precisely simulate even complex optical systems such as camera lenses. By means of the software package *LightTools* [Syn], sophisticated illumination systems can be designed, optimized, and simulated.

2.3 Summary

This chapter provided a thorough review of the current state of the art, divided into the science and the technology, of the topics covered by this thesis. There are several mature scientific and technological solutions for the visual inspection of transparent objects and the different components involved in this task. However, most of these systems and products are very specialized, for a certain type of test object, e.g., for flat glasses, or for bottles, or for contact lenses. There is currently no published inspection approach that is applicable to complex-shaped objects without the need for a time-consuming adaptation of the system components to the specific type of test object.

This thesis contributes to the state of the art in all the subjects covered in this chapter. The major contributions are:

- An optical concept of a fully programmable light field generator including an adequate calibration procedure (see Section 3).
- To precisely measure light deflection angles with a resolution that is constant over the whole measurement field, a novel optical setup of a $4f$-light field camera is described (see Section 4.2.2) and simulated.

- A laser deflection scanner is introduced (see Section 4.4.1) that acquires the light field information of large and complex-shaped transparent test objects.
- An optical setup to acquire the light transport matrices of transparent objects (see Section 4.5) and methods to process these transport matrices to visualize any material defects (see Section 6.4).
- A thorough set of additional plugins for the rendering framework Mitsuba, allowing the simulation of complete machine vision setups for visual inspection applications (see Chapter 5).
- A framework for processing the light fields of transparent test objects in order to visualize absorbing and scattering material defects (see Chapter 6).
- An adaptation and fast implementation of the Generalized Cramér-von Mises Distance to be used as a distance function in the processing framework mentioned above.
- Image fusion algorithms that optimize a specifically formulated signal-to-noise ratio for the inspection of transparent objects. By exploiting information about the test object's nominal state, the fusion methods highlight any material defects while suppressing the object's geometrical features (see Section 6.3).
- A visual inspection method that visualizes material defects by illuminating a transparent test object with an inverse light field. By this means, the object's nominal appearance is suppressed and only deviations from the intended state remain visible (see Chapter 7).

These contributions pave the way for the development of inspection systems that can easily be adapted to new kinds of transparent test objects and are robust regarding complex geometries.

3 Light field illumination

The illumination component is an important part of machine vision and computer vision systems. Established light sources allow adjusting the intensity and, to a certain degree, the spectrum of the emitted light. For displays, these parameters can even be independently set for different spatial positions, i.e., the single pixels. However, the emitted light's direction of propagation is usually fixed and cannot be controlled by the user. In contrast, light field generators are able to also vary the angle of the emitted light rays. Their optical setup relies either on the concept of holography, stacked transmissive masks, grating technologies, or microlens arrays. This last technique forms the basis of the light field generator introduced in this thesis. The optical setup of the introduced light field generator is described in Section 3.1 and Section 3.2 covers its calibration. The developed prototype is described in Section 3.3 and results of experiments evaluating its calibration are presented in Section 3.4. Section 3.5 closes the chapter with a summary.

3.1 Optical setup

The proposed light field generator is based on a combination of a high-resolution two-dimensional display, i.e., a spatially programmable light source, and a microlens array. The microlens array is aligned parallel to the display at a distance equal to that of the microlenses' focal length f_{ML}. By this means, a collimated beam of light, i.e., a bundle of light rays propagating parallel to each other, is emit-

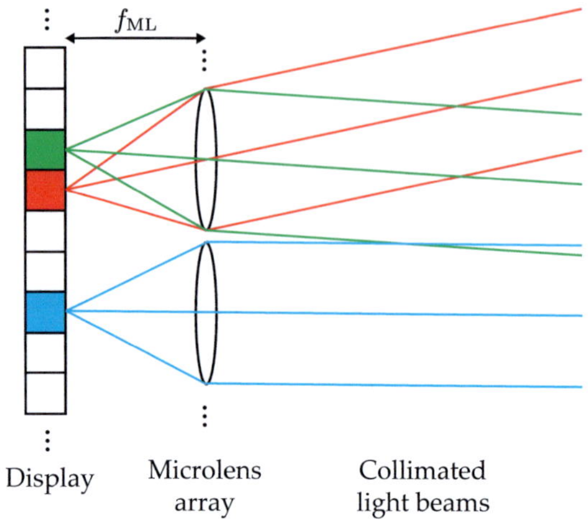

Figure 3.1: Basic concept of the light field generator based on a microlens array. The light emitted by the display's pixels is transformed into collimated light beams by the lenses of the microlens array (example optical paths in red, green and blue). To simplify matters, the pixels are treated as point light sources.

ted if a display pixel located under the respective microlens is turned on (see Fig. 3.1). Hence, the spatial resolution of this light field generator is reduced to the resolution of the microlens array. The collimated beam originates from every microlens located inside the angle of emission of the activated pixel. The angle α of an emitted collimated light bundle depends on the focal length f_{ML} and on the distance δ_α of the active pixel to the optical axis of the corresponding microlens (for an illustration, see Fig. 3.2):

$$\tan(\alpha) = \frac{\delta_\alpha}{f_{\mathrm{ML}}}\,. \tag{3.1}$$

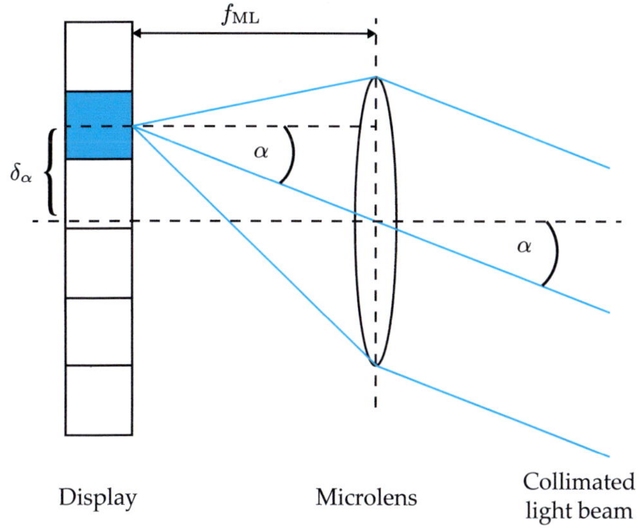

Figure 3.2: Geometric relations between the microlenses' focal length f_{ML}, the relative position δ_α of the active pixel, and the angle α of the corresponding collimated light beam.

Important parameters of a light field generator are

- the spatial resolution, i.e., the numbers M_{ML} and N_{ML} of spatial elements (lenses) in the horizontal, respectively, the vertical direction, and their respective pitches s_{h} and s_{v},

- the total ranges of possible angles θ_{t} and φ_{t} of emitted collimated light beams in the horizontal, respectively, vertical direction, and

- the angular resolution, i.e., the numbers A_{h} and A_{v} of independent emission angles in the horizontal, respectively, vertical direction, and their respective pitches θ_Δ and φ_Δ.

The spatial resolution M_{ML} and N_{ML} is given by the number of microlenses in the horizontal, respectively, the vertical direction. The

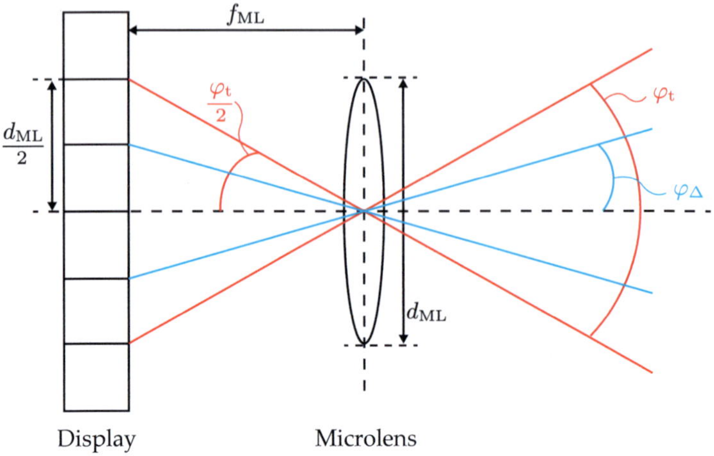

Figure 3.3: The total possible vertical emission angle φ_t and the vertical angular pitch φ_Δ of a light field generator.

pitches s^{h} and s^{v} between the centers of the microlenses in the horizontal, respectively, vertical direction, is determined by the geometric layout of the microlens array. The total range of possible emission angles θ_t and φ_t depends on the microlenses' focal length f_{ML} and on their diameter d_{ML}. As illustrated in Fig. 3.3, these angles[1] can be calculated by means of the triangle defined by f_{ML}, d_{ML} and φ_t via

$$\tan\left(\frac{\varphi_t}{2}\right) = \frac{d_{\mathrm{ML}}}{2} \cdot \frac{1}{f_{\mathrm{ML}}} \Leftrightarrow \varphi_t = 2 \cdot \tan^{-1}\left(\frac{d_{\mathrm{ML}}}{2} \cdot \frac{1}{f_{\mathrm{ML}}}\right), \quad (3.2)$$

with $\tan^{-1}$ denoting the inverse tangent function. The angular resolution is given by the number of display pixels under one microlens, i.e., by

$$A_{\mathrm{h}} = \frac{d_{\mathrm{ML}}}{e_{\mathrm{h}}}, \quad A_{\mathrm{v}} = \frac{d_{\mathrm{ML}}}{e_{\mathrm{v}}}, \quad (3.3)$$

[1] Here, the case of circular or square lenses is considered, hence $\theta_t = \varphi_t$.

with e_h and e_v denoting the horizontal, respectively, vertical sizes of the pixel elements of the display. The angular pitches θ_Δ and φ_Δ can be estimated via

$$\theta_\Delta \approx \frac{\theta_\mathrm{t}}{A_\mathrm{h}}, \quad \varphi_\Delta \approx \frac{\varphi_\mathrm{t}}{A_\mathrm{v}}. \tag{3.4}$$

As indicated by the symbol $\approx$, these equations are only approximations, due to the non-linear nature of the tangent function (the angular pitches decrease slightly with increasing distance of the emitting pixel from the optical axis of the considered microlens). Indeed, for the dimensions involved in the case of a light field generator, the errors introduced by these approximations can be neglected.

3.2 Calibration

The purpose of a light field generator is to emit $L(x, y, \theta, \varphi)$, a user-defined light field.[2] The (x, y)-coordinates are relating to the plane of the light field generator where rays contained in L will originate from. Given a light field L, the introduced light field generator has to determine which pixels of its display to turn on in order to emit a light field that matches L as well as possible. Hence, a calibration of the device is required, which yields a function

$$f_\mathrm{LF} : (x, y, \theta, \varphi) \mapsto \mathbf{u}, \tag{3.5}$$

mapping a requested ray (x, y, θ, φ), for which $L(x, y, \theta, \varphi) > 0$, to a pixel $\mathbf{u} = (u, v)^\mathsf{T}$, $u \in [1, \ldots, U]$, $v \in [1, \ldots, V]$ of the display of the light field generator whose corresponding collimated light bundle

[2] In the scope of this thesis, the dependence of the light field on the wavelengths λ of the contained rays is omitted. If the display employed in the light field generator supports emitting different spectra, λ can be added as an additional parameter to the definition of L.

provides the best possible approximation of this ray. In the calibration procedure, for each microlens the central pixel $\widetilde{\mathbf{u}}$, i.e., the pixel corresponding to a light bundle propagating parallel to the optical axis of the corresponding microlens, is determined. To emit a requested ray bundle (x, y, θ, φ), the central pixel $\widetilde{\mathbf{u}}_{(x,y)}$ of the microlens corresponding to (x, y) is selected first. Then, the emission angle (θ, φ) is translated into the pixel offset $\Delta\mathbf{u}_{(\theta,\varphi)}$ with respect to $\widetilde{\mathbf{u}}_{(x,y)}$ by calculating the corresponding spatial displacement on the plane of the display via Eq. (3.1). By this means, f_{LF} can be evaluated:

$$f_{\mathrm{LF}}(x, y, \theta, \varphi) = \widetilde{\mathbf{u}}_{(x,y)} + \Delta\mathbf{u}_{(\theta,\varphi)} \,. \tag{3.6}$$

Determination of central pixels A calibration setup is arranged consisting of the light field generator, an industrial monochrome camera equipped with a telecentric lens, and an image processing computer. The optical axes of the microlenses of the light field generator and of the telecentric lens are matched as closely as possible. Due to the telecentric lens, the camera will only capture those light bundles emitted by the light field generator that propagate parallel to the optical axis. As explained before, light bundles directed parallel to the optical axis are generated if the central pixel under a microlens is turned on. By means of two binary coded illumination series, the set $\mathcal{L}$ of all display pixels representing central pixels can be efficiently determined as it is explained in the following.

To acquire the set $\mathcal{L}$ of central pixels, the light field generator displays illumination patterns that successively encode the coordinates of the respective pixels. The first series encodes the horizontal u-coordinate and the second series encodes the vertical v-coordinate. In every illumination step, one bit of the coordinate of every pixel is encoded and transmitted to the receiving camera. Algorithm 3.1 lists the corresponding pseudocode.

Algorithm 3.1: Acquisition of the set $\mathcal{L}$ of central pixels of a light field generator by means of a coded illumination series.

```
function ACQUIRECENTRALPIXELS(camera, LFGenerator)
    g_0(m) ← camera.acquireImage()
    L ← ∅
    g_hor(m) ← emptyImage(M, N)
    g_ver(m) ← emptyImage(M, N)
    for i ∈ [1, ..., ⌈log_2(U)⌉] do
        U ← {u ∈ [1, ..., U] : bin(u)_i = 1}
        V ← [1, ..., V]
        LFGenerator.turnOnPixels(U × V)
        g(m) ← camera.acquireImage() − g_0(m)
        bin(g_hor(m))_i = { 1  if g(m) > t,
                          { 0  otherwise
    end for
    for i ∈ [1, ..., ⌈log_2(V)⌉] do
        U ← [1, ..., U]
        V ← {v ∈ [1, ..., V] : bin(v)_i = 1}
        LFGenerator.turnOnPixels(U × V)
        g(m) ← camera.acquireImage() − g_0(m)
        bin(g_ver(m))_i = { 1  if g(m) > t,
                          { 0  otherwise
    end for
    L = {(g_hor(m), g_ver(m))^T , m ∈ [1, ..., M] × [1, ..., N]}
    return L
end function
```

This approach uses the binary representation of the u- and v-coordinates of the pixels of the light field generator. In this context, $\mathrm{bin}(u)_i$ refers to the i-th bit of the binary representation of u beginning at the most significant bit. Initially, a reference image $g_0(\mathbf{m})$, $\mathbf{m} = (m, n)^\intercal$ is acquired while the light field generator `LFGenerator` is not emitting any light, in order to capture the influence of the surroundings. Two empty images $g_{\mathrm{hor}}(\mathbf{m})$ and $g_{\mathrm{ver}}(\mathbf{m})$ of size $M \times N$, which is the size of the images captured by the employed camera, are allocated. They will store the decoded positions of the light field generator's pixels received by the camera.

The first for-loop implements the acquisition of the u-coordinates. In iteration i, all pixels with $\mathrm{bin}(u)_i = 1$ are turned on and the camera captures an image from which $g_0(\mathbf{m})$ is subtracted in order to remove the influence of any background illumination sources. If the resulting $g(\mathbf{m})$ is greater than a threshold t, this is an indication that a pixel of the light field generator, with its horizontal coordinate having the i-th bit set to one, has been imaged to the camera pixel $\mathbf{m}$. Hence, the corresponding bit $\mathrm{bin}(g_{\mathrm{hor}}(\mathbf{m}))_i$ of the result image for the horizontal coordinate can be set to one—otherwise it is set to zero.

Similarly, the v-coordinates are acquired by means of a second for-loop. The images $g_{\mathrm{hor}}(\mathbf{m})$ and $g_{\mathrm{ver}}(\mathbf{m})$ now contain the u- and v-coordinates of the pixels of the light field generator that have been imaged to the camera pixel $\mathbf{m}$. They are now combined into the sought set of central pixels $\mathcal{L}$. The presented algorithm has a complexity[3] of $\mathcal{O}(\lceil \log_2(U) \rceil + \lceil \log_2(V) \rceil)$.

To assign the found central pixels to the individual microlenses, the image region of the light field generator in the image captured by the telecentric camera is divided into square cells of the same size as the microlenses. Because the grid periods of the microlens array and the display pixels are not necessarily integer multiples, it is possible that there are several or zero central pixels for some microlenses as it is illustrated by Fig. 3.4. In the case of multiple brightest central

[3] A short description of the so-called *Big O notation* for expressing computational complexities is given in Appendix A.1.

pixels, one of them is chosen at random. If there is no center pixel, queries of $f_{\mathrm{LF}}(x, y, \theta, \varphi)$ regarding the affected microlenses are discarded.

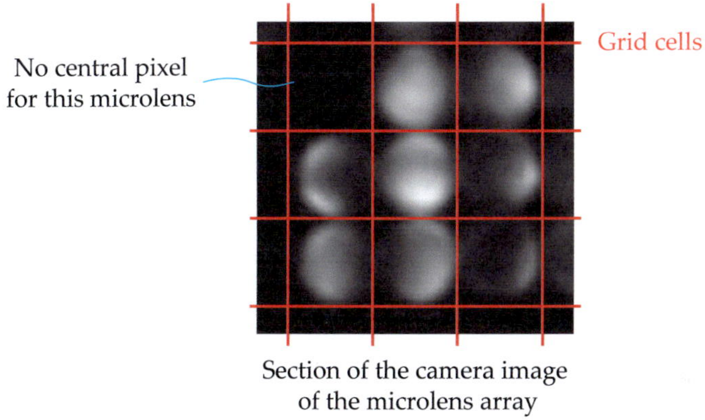

Figure 3.4: The grid of square cells into which the image of the light field generator is divided. The image shows a section of 3×3 microlenses for which the determined central pixels are turned on. Since not all of the central pixels are located directly under the center of their corresponding microlens, the microlenses appear with different brightnesses or are even completely black (upper left microlens).

3.3 Light field generator prototype

For an initial prototype, the Sony Xperia Z5 Premium smartphone was chosen as the display device because of its high spatial resolution of 2160×3840 pixels. It has been combined with a 100×100 microlens array with $f_{\mathrm{ML}} = 3000\ \mu\mathrm{m}$ and $d_{\mathrm{ML}} = 645\ \mu\mathrm{m}$ manufactured by Fraunhofer-IOF (Germany). The result has a spatial resolution of 100×100, with a vertical and horizontal pitch of $s_{\mathrm{h}} = s_{\mathrm{v}} = 650\ \mu\mathrm{m}$, total ranges of possible angles of $\theta_{\mathrm{t}} = \varphi_{\mathrm{t}} \approx 12.27°$, angular resolution of $A_{\mathrm{h}} = A_{\mathrm{v}} = 10$ and corresponding angular pitches of $\theta_{\Delta} = \varphi_{\Delta} \approx 1.23°$, since approximately 10×10 display pixels fit un-

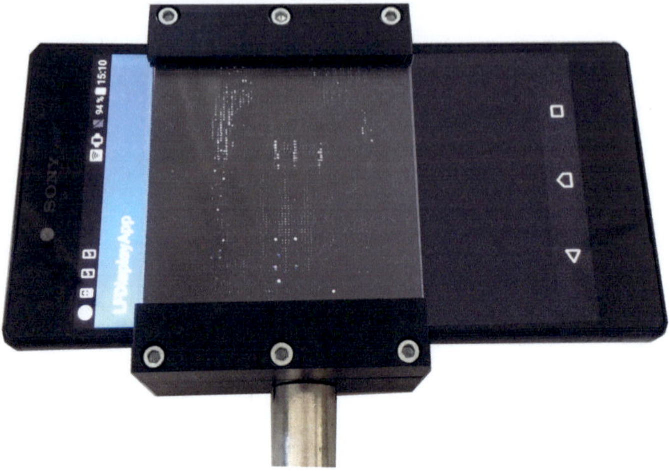

Figure 3.5: Photograph of the prototype of the light field generator emitting the inverse field used in the experiments presented in Section 7.2.2.3. The microlens array is held by mechanical mounts with a distance of f_{ML} to the plane of the display. By means of the rod at the bottom, the light field generator can be mounted on an optical bench to conduct experiments.

der one microlens.[4] Figure 3.5 shows a photograph of the prototype emitting a light field for the experiments with the inverse light field illumination method (see Section 7.2.2).

For controlling the light field generator, a suitable app (short for application) has been developed and installed on the Android powered smartphone. To allow an efficient and platform independent communication with the device, a network protocol based on TCP (transmission control protocol) has been formulated for transferring a set of pixel coordinates corresponding to the requested light field from a client device to the light field generator. The communication

[4] According to the technical specifications of the Sony Xperia Z5 Premium, about 20 × 20 display pixels should fit under one microlens. However, the full resolution of the device is only available under certain circumstances since it employs a sophisticated approach for multiplexing the red, green or blue sub-pixels of adjacent pixels.

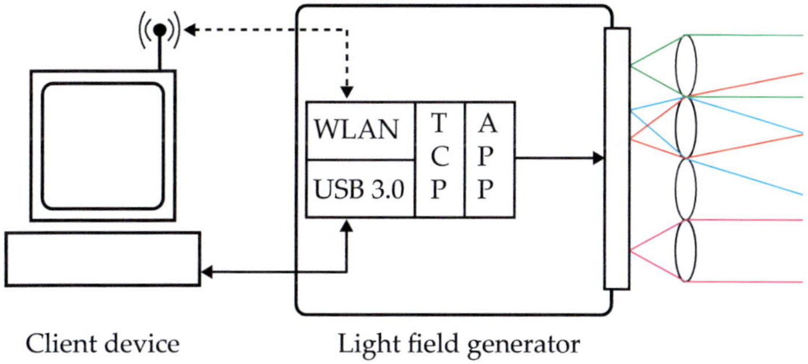

Figure 3.6: Communication concept for controlling the developed light field generator. Via a network protocol, a client device can transmit the requested light field to the app controlling the display of the light field generator.

concept is illustrated by Fig. 3.6. Since the employed smartphone can be connected to other devices via WLAN (wireless local area network) or USB 3.0 (universal serial bus), the light field generator can be easily integrated into an experiment.

3.4 Experiments

The developed prototype has been calibrated according to the calibration procedure described in Section 3.2. To validate the calibration method, images have been acquired with a telecentric camera for different requested light fields.[5] Figure 3.7 shows the resulting camera images. The top left-hand image shows the camera image resulting when a light field is requested, where all microlenses are supposed to emit rays parallel to their optical axis, i.e., when those pixels of the light field generator are turned on, which have been determined in the calibration procedure. This corresponds to emission angles of $(\theta, \varphi) = (0^\circ, 0^\circ)$. Many of the microlenses appear bright as

[5] More details about the used components are provided in Appendix A.3.

expected. The dark gaps are caused by microlenses where no central pixel could be determined.

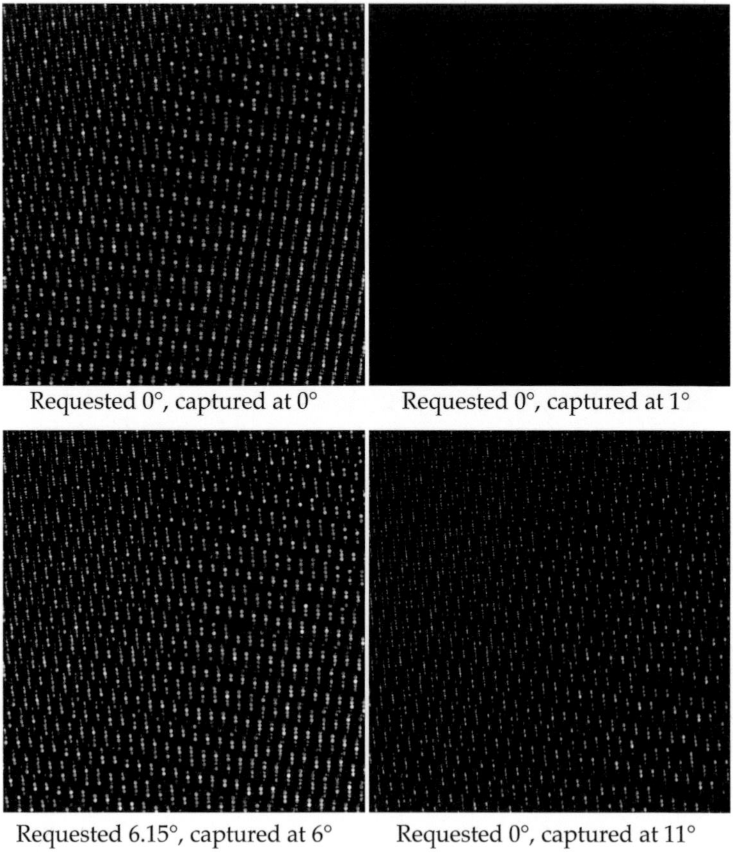

Figure 3.7: Results of the experiments performed after the calibration of the prototype of the light field generator. All images share the same intensity range.

The top right-hand image resulted when requesting the same light field but rotating the light field generator by $1°$ around its vertical axis. As can be seen, there are no light rays directed parallel to the optical axis of the telecentric camera. This supports the theoretically determined horizontal angular pitch of $\theta_\Delta \approx 1.23°$.

For the bottom left-hand image, the horizontal u-coordinate of the previously displayed pixels of the light field generator have been increased by 5. This corresponds to emission angles of

$$(\theta, \varphi) = (0° + 5\theta_\Delta, 0°) = (6.15°, 0°)\,. \tag{3.7}$$

The shown image has been acquired for a rotation angle of about $6°$ around the vertical axis of the light field generator. Since approximately as many microlenses appear bright as for the first image, this result also supports the chosen parametrization of the light field generator and the presented calibration procedure.

The bottom right-hand image corresponds to emission angles of $(\theta, \varphi) = (0°, 0°)$ and to a rotation of the light field generator by approximately $11°$ around its vertical axis. Although these emission angles were not requested, corresponding light rays are emitted. This observation can be explained with the phenomenon of so-called *crosstalk*-effects. Since the space between the microlenses is not opaque, display pixels can also lead to collimated light beams originating from adjacent microlenses with angles larger than $(\theta_\mathrm{t}, \varphi_\mathrm{t})$. This drawback could be mitigated by mounting a grid of opaque walls between the display and the microlens array. By this means, a microlens could only be illuminated by the pixels inside its projection on the display plane.

3.5 Summary

This chapter introduced an optical setup for the realization of a fully programmable light field generator. By joining a microlens array with a high-resolution two-dimensional display, light beams with individual spatial origins and propagation directions can be emitted. Furthermore, a calibration procedure has been formulated, which allows obtaining the display pixel corresponding to the spatial location and emission angle of a desired light beam.

An initial prototype based on a high-resolution smartphone display and a 100×100 microlens array has been built. An application and a network protocol have been developed, that allow an efficient and flexible communication with the device to control the light field generator. Performed experiments proved that the presented concept is basically suitable for realizing a light field generator and that the calibration can be conducted according to the introduced procedure. The current prototype still has some limitations as not all emission directions are possible due to the discrete grid structure of the display pixels and the microlens array. However, this issue can be resolved by manufacturing a microlens array that better fits to the employed display. Furthermore, the effect of crosstalk could be mitigated by inserting opaque walls separating the groups of pixels corresponding to individual microlenses.

4 Light field acquisition

Light field images can be acquired with several approaches. As pointed out in the introduction, light fields should contain enough information to test transparent objects for both absorbing and scattering material defects. For the acquisition of light fields, this chapter introduces light field cameras in Section 4.2, methods based on schlieren imaging in Section 4.3, and the concept of laser deflection scanners in Section 4.4. In the context of the visual inspection of transparent objects, the obtained light fields are sometimes referred to as *light deflection maps*, indicating a collimated illumination of the test objects. The concept of light deflection maps is introduced in Section 4.1. Besides these light deflection maps, also light transport matrices can be acquired and processed, as they also contain information about the direction of propagation of the light. Section 4.5 presents a method for the acquisition of light transport matrices. The data acquired by the different approaches described in the following sections is processed by the processing methods presented in Chapter 6.

4.1 Light deflection maps

Light deflection maps contain spatially resolved information about the angular distribution of the propagation directions of collimated light rays after being transmitted through the measurement field, e.g., a transparent test object. Figure 4.1 illustrates the principle concept of the acquisition of light deflection maps and Fig. 4.2 shows real light deflection maps of a cylindrical lens acquired with a laser deflection scanner.

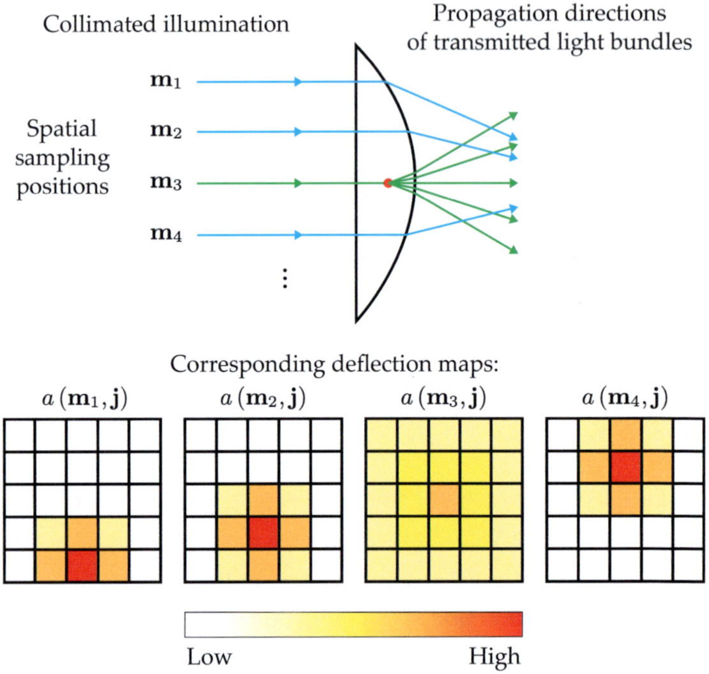

Figure 4.1: Visualization of the acquisition of light deflection maps. The test object, a plano-convex lens, is illuminated with parallel light bundles. Each light bundle represents an individual spatial sampling position. The light deflection maps show the angular distribution of the light bundles' directions of propagation in pseudo-colors after being transmitted through the test object. The blue rays are refracted at the test object's surface, which is why the corresponding deflection maps have comparatively concentrated peaks. The green ray is scattered in multiple directions by a scattering material defect, resulting in a deflection map with a broader intensity distribution instead of a distinct peak.

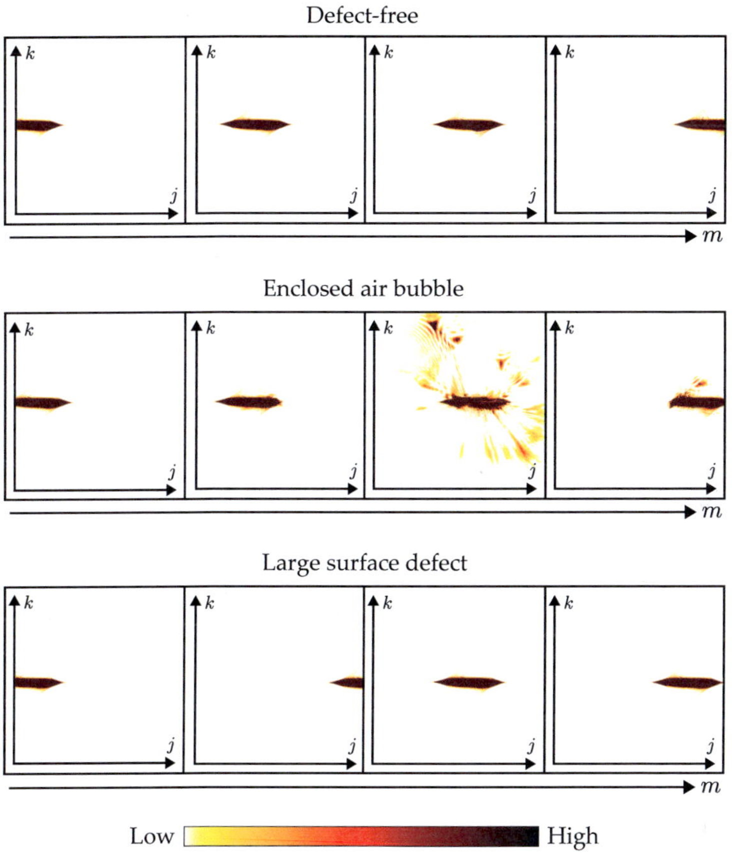

Figure 4.2: Example deflection maps of a cylindrical lens. For defect-free regions of the test object, the peaks (oblong shaped due to the cylindrical lens) are concentrated and their positions correspond to the deflections caused by the slope of the test object's surface (first row). An enclosed air bubble scatters the incident light and leads to a broad distribution of the deflection directions (second row, third image). A larger surface defect leads to the corresponding peak being noticeably shifted with respect to its intended position (third row, second image).

A light deflection map (from now on called a *deflection map*) is a four-dimensional structure

$$a(\mathbf{m},\mathbf{j}),\ \mathbf{m}=(m,n)^{\mathsf{T}},\mathbf{j}=(j,k)^{\mathsf{T}}\,, \tag{4.1}$$

$$(m,n)\in\Omega_{\mathrm{s}}=\{1,2,\ldots,M\}\times\{1,2,\ldots,N\}\subset\mathbb{Z}^2 \tag{4.2}$$

$$(j,k)\in\Omega_{\mathrm{a}}=\{1,2,\ldots,J\}\times\{1,2,\ldots,K\}\subset\mathbb{Z}^2\,, \tag{4.3}$$

composed of two discrete spatial coordinates $(m,n)^{\mathsf{T}}$ and two discrete angular coordinates $(j,k)^{\mathsf{T}}$ with a spatial domain of Ω_{s} and an angular domain of Ω_{a}. Figure 4.3 illustrates the structural concept of a deflection map. The spatial coordinates result from the spatial sampling performed by the acquisition device. The angular coordinates contain the two-dimensional deflection angles of the collimated light rays after being transmitted through the measurement field. Since all presented sensors have a discrete detector structure, both $\mathbf{m}$ and $\mathbf{j}$ are discrete quantities. For many applications, the employed sensors yield square deflection distributions, i.e., $J=K$. The transformation of the light field $L(x,y,\theta,\varphi)$ exiting the test object to the corresponding deflection map $a(\mathbf{m},\mathbf{j})$ can also involve non-linear operations (e.g., the tangent of the deflection angle). In addition, depending on the acquisition method, the deflection maps can have a slightly different interpretation. For a light field camera, the spatial sampling positions are linked to the object point where the light rays exit the test object and not to the lateral positions of the light rays inside the collimated beam as shown in Fig. 4.1. Further details will be pointed out in the corresponding sections.

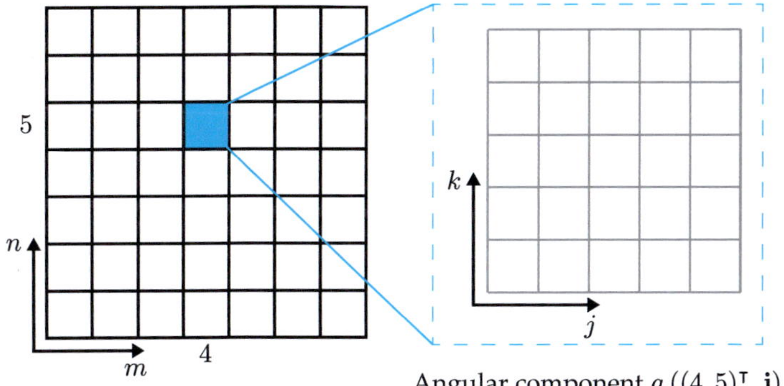

Figure 4.3: Concept of a deflection map: For every spatial position $\mathbf{m}$ sampled by the acquisition device, the deflection map $a(\mathbf{m}, \mathbf{j})$ contains the two-dimensional distribution of the captured deflection angles, with the coordinate $\mathbf{j}$ corresponding to the angular component.

4.2 Camera-based light field imaging

In contrast to conventional cameras, light field cameras also capture the direction of the incident light rays and by this means they acquire a four-dimensional section of the incident light field. Since their basic optical setup is based on that of conventional cameras, the following paragraph will provide an introduction to the basic thin lens camera.

Thin lens camera The thin lens camera model allows describing the image formation process for an optical system consisting of an image sensor and a lens (system) for which the assumptions of geometric optics apply. Figure 4.4 shows the principle of the optical setup of a thin lens camera. In general, three different coordinate systems are used when describing the image formation of a camera system: the world coordinate system $(x^{\mathrm{w}}, y^{\mathrm{w}}, z^{\mathrm{w}})$, which is located in the observed scene, the camera coordinate system $(x^{\mathrm{c}}, y^{\mathrm{c}}, z^{\mathrm{c}})$, which is attached to the camera[1], and the image coordinate system $(x^{\mathrm{b}}, y^{\mathrm{b}})$ located in the sensor plane. Since discriminating between the world coordinate system and the camera coordinate system is of no special importance for the following content, it is assumed that they both coincide and only the camera coordinate system is used. Points of the observed scene are called object points and are denoted by $\mathbf{p}^{\mathrm{c}} = (p_x^{\mathrm{c}}, p_y^{\mathrm{c}}, p_z^{\mathrm{c}})$. The lens is characterized by its focal length f and by its diameter d. The distance from the sensor to the lens is referred to as the image distance and is denoted by b. Light bundles diverging from the observed scene are transformed into converging light bundles by the lens. If the object point $\mathbf{p}^{\mathrm{c}}$ from which the light bundle originates is located in the plane of focus of the system, the bundle converges on the sensor plane at the image point $\mathbf{p}^{\mathrm{b}} = (p_x^{\mathrm{b}}, p_y^{\mathrm{b}})$, leading to a sharp image of the object point. This coordinate transformation is called a central projection [Hec17]. The correspondence between an object point $\mathbf{p}^{\mathrm{c}}$ and its corresponding image point $\mathbf{p}^{\mathrm{b}}$ can be geometrically determined by means of the so-called *principal ray* and the *parallel rays* as illustrated in Fig. 4.4: the dotted green line represents the principal ray connecting $\mathbf{p}^{\mathrm{c}}$ and $\mathbf{p}^{\mathrm{b}}$ and running through the center of the lens. The parallel rays (dotted red lines) run parallel to the optical axis on one side of the lens and cross the optical axis on the other side of the lens at its focal point, i.e., at a distance of f from the lens. The principal ray intersects with the parallel rays at $\mathbf{p}^{\mathrm{c}}$ and $\mathbf{p}^{\mathrm{b}}$. The plane of focus is located parallel to the $(x^{\mathrm{c}}, y^{\mathrm{c}})$-plane of the camera coordinate system, at a distance of g. The relation between g,

[1] As a commonly used convention, the camera's line of sight is assumed to coincide with the z-axis z^{c} of the camera coordinate system.

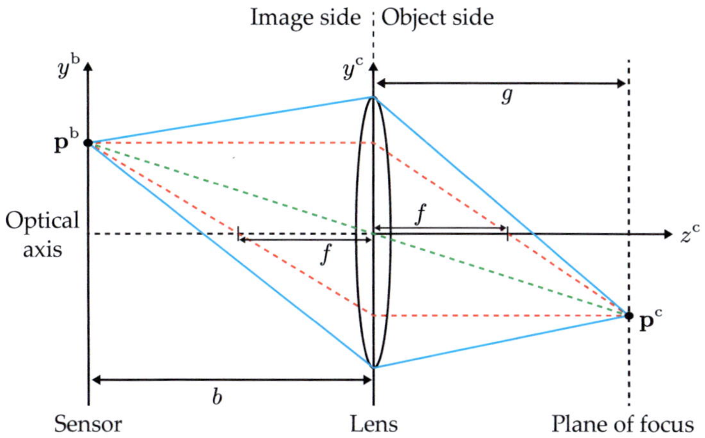

Figure 4.4: The optical principle of the thin lens camera model. An object point $\mathbf{p}^c$ is focused on the sensor plane at the image point $\mathbf{p}^b$. From the light bundle diverging from $\mathbf{p}^c$, the lens captures the section drawn in blue. The dotted red lines are the parallel rays and the dotted green line is the principal ray. The z^c-axis originates at the intersection of the optical axis and the lens. The x^b-axis originates at the intersection of the optical axis and the sensor plane and the x^c-axis originates at the intersection of the y^c-axis and the z^c-axis. Both the x^b-axis and the x^c-axis are aligned perpendicularly to the drawing plane and point upwards.

the lens's focal length f, and the image distance b, is given by the so-called thin lens formula [Hec17]:

$$\frac{1}{f} = \frac{1}{g} + \frac{1}{b}\,. \tag{4.4}$$

The relation between the real height of the imaged object in the scene to the height of its image on the sensor is called the magnification

$$V = \frac{|p_y^b|}{|p_y^c|} \tag{4.5}$$

of the optical system. The thin lens formula can be derived by means of the intercept theorem [Agr08]. As depicted in Fig. 4.4, all light rays of the captured light bundle (blue cone) converge on the sensor at $\mathbf{p}^{\mathrm{b}}$ and contribute to the measured electrical signal. For a constant light intensity, the measured signal and therefore also the signal-to-noise ratio increase for a larger lens diameter d_{L} since this increases the captured light bundle.[2]

Acquisition of deflection angles Besides making it possible to obtain a point to point imaging of an observed scene, a lens also allows acquiring information about the light's direction of propagation in terms of the angles through which the light rays are deflected with respect to the optical axis of the lens. Figure 4.4 shows that the light rays intersect the focal plane of the lens on the image side at distances to the optical axis that are related to the directions of propagation of the rays on the object side. When imaging an object point at *optical infinity*, i.e., at a distance $g \gg f$, the light rays captured by the lens form a so-called *collimated* light bundle, i.e., they propagate approximately parallel to each other. Therefore, these rays intersect on the image-side focal plane of the lens. As is illustrated by Fig. 4.5, the distance δ_α between the intersection of the ray bundle and the optical axis can be determined by means of straightforward geometric considerations. Since the focal length f, the angle α by which the collimated light bundle is tilted with respect to the optical axis, and the sought distance δ_α define a right triangle, it follows that

$$\delta_\alpha = f \cdot \tan(\alpha) \ . \tag{4.6}$$

[2] A larger lens diameter (and a shorter focal length) also decreases the depth of field, i.e., how far the imaged object can be moved away from the plane of focus without leading to a blurred image. Further information can be found in [Bey15].

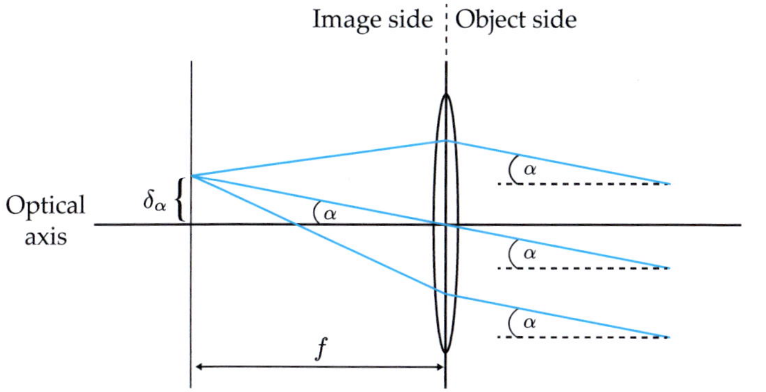

Figure 4.5: Transformation of angular information about the propagation direction of light into spatial information by a lens. An incident collimated light bundle tilted by α with respect to the optical axis is imaged to a point with a distance δ_α to the optical axis.

The signs of the coordinates of the intersection with the focal plane are determined by the orientation of the collimated light bundle.

Hence, a lens transforms angular information about the direction of propagation of the light on its object side to spatial information on the focal plane on its image side, and vice versa. By placing a sensor in the focal plane on the image side of the lens, information about the angular distribution of the direction of propagation of the captured light can be acquired. However, any spatial information is lost and has to be obtained by other means. Sampling the intensity distribution on the focal plane is a key principle in the field of light field imaging, which is used by most of the optical setups described in the following sections for the acquisition of light fields.

4.2.1 Light field camera

Based on the model of the thin lens camera, the concept of a light field camera can now be explained. Figure 4.6 shows the principle of the optical setup of a light field camera. In such a camera, a section of the four-dimensional light field $L(x, y, \theta, \varphi)$ is spatially multiplexed on a two-dimensional sensor. To this end, a microlens array, i.e., a two-dimensional grid of small lenses, is positioned at the image plane of the main lens, i.e., at a distance of b_{L}, where the sensor is usually mounted in a conventional camera (cf. Fig. 4.4). The sensor is placed behind the microlens array at a distance b_{ML} of the focal length f_{ML} of the microlens array, hence $b_{\mathrm{ML}} = f_{\mathrm{ML}}$. A divergent ray bundle originating from a point on the plane of focus in the observed scene is transformed by the main lens into a ray bundle that converges on the plane of the microlenses. The converging light rays are split up by the corresponding microlens and reach different pixels of the sensor, depending on their direction of propagation in the scene. This setup is a realization of the principle of measuring angular information at the image-side focal plane of a lens, as introduced before (see p. 64). All pixels which correspond to a certain microlens, i.e., which are located inside the projected area of the microlens on the sensor, observe the same object point, but from different directions. Ideally, the rays split up by one microlens cover the whole corresponding sensor area but do not lead to crosstalk, i.e., they do not reach the sensor area of neighboring microlenses. This is the case provided

$$\frac{d_{\mathrm{ML}}}{b_{\mathrm{ML}}} = \frac{d_{\mathrm{L}}}{b_{\mathrm{L}}}, \tag{4.7}$$

with d_{ML} denoting the diameter of one microlens, and d_{L} denoting that of the main lens [Ng05]. Since the sensor is located in the focal plane of the microlenses, this part of the optical system is focused

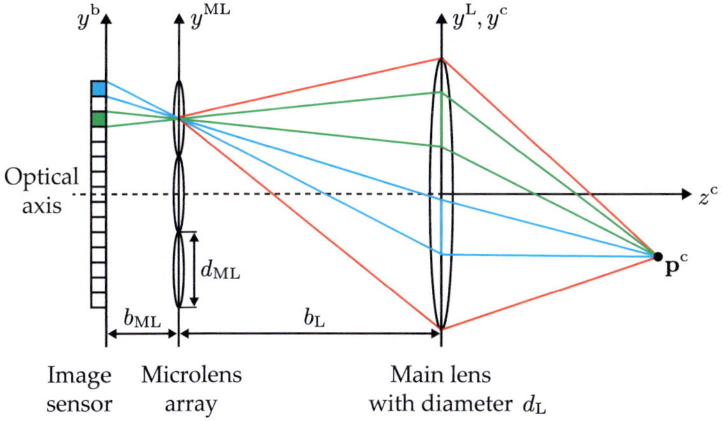

Figure 4.6: Principle of the optical setup of a light field camera. The main lens realizes a sharp optical image of the object point $\mathbf{p}^{\mathrm{c}}$ onto the plane of the microlens array. The microlens array images the plane of the main lens onto the sensor plane. Hence, the blue pixel observes only the blue section of the whole light bundle (drawn in red) and the green pixel observes only the green section. All the pixels of the top microlens observe the same object point, but from different directions.

at optical infinity. As the distance b_{L} between the microlenses and the main lens is very large compared to f_{ML}, the microlenses yield a sharp image of the plane of the main lens on the image sensor. Consequently, every pixel under one microlens observes a section of this plane, i.e., a sub-aperture, and integrates all the light rays with the corresponding propagation directions (e.g., the green and blue cones in Fig. 4.6).

A common way to parametrize the captured light field L' is as $L'\left(x^{\mathrm{ML}}, y^{\mathrm{ML}}, x^{\mathrm{L}}, y^{\mathrm{L}}\right)$ with pairs of points $\left(x^{\mathrm{ML}}, y^{\mathrm{ML}}\right)^{\intercal}, \left(x^{\mathrm{L}}, y^{\mathrm{L}}\right)^{\intercal}$ located on the plane of the microlens array, respectively, on the plane of the main lens. Another parametrization, which is closely linked to the introduced deflection maps, is given by $L'(m, n, j, k)$, with $(m, n)^{\intercal}$ representing a microlens (in the m-th row and the n-th column) and $(j,k)^\intercal$ addressing one of the pixels under the respective microlens (cf. Fig. 4.3). In order to obtain a deflection map of a transparent test object as introduced in Section 4.1 with such a light field camera, the object has to be illuminated with a collimated light beam. However, the captured $L'(m,n,j,k)$ cannot be directly used as a deflection map, since the directions of propagation represented by a relative pixel position $(j,k)^\intercal$ under a microlens differ across the microlens array, as illustrated in Fig. 4.7. In order to mitigate this problem, the introduced optical setup of a light field camera can be extended to a $4f$-light field camera, which is introduced in Section 4.2.2. Due to these limitations of a conventional light field camera, all the light field experiments described in Section 6.1.3.1 based on light field camera images use simulated light field images acquired with a model of the $4f$-light field camera.

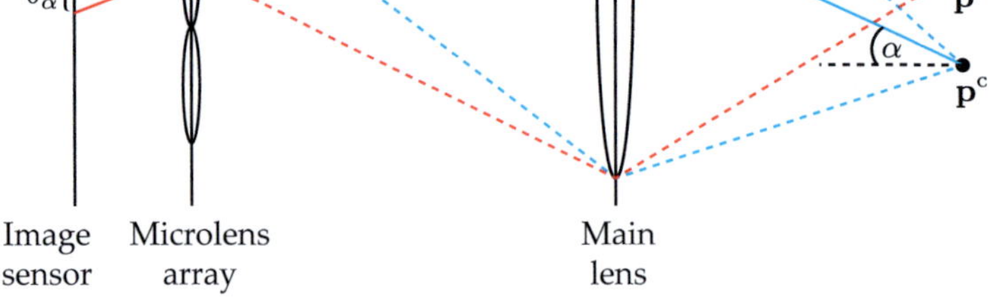

Figure 4.7: At the two object points $\mathbf{p}^c$ and $\mathbf{p}'^c$, the incident light rays (blue and red) are deflected upwards by the angle α with respect to a line parallel to the optical axis. Although both rays have the same deflection angle, they hit the sensor with different displacements $\delta_\alpha \neq \delta'_\alpha$ with respect to the projected center of the corresponding microlenses.

4.2.2 $4f$-light field camera

By adding a second lens to the optical setup of the light field camera described above, a so-called *$4f$-light field camera* can be realized. Figure 4.8 shows the corresponding optical concept. Similar to the conventional light field camera, the image sensor is placed in the focal plane of the microlens array. The first lens is positioned at a distance equal to its focal length f_{L1} from the microlens array and has a distance of $f_{L1} + f_{L2}$ from the second lens with focal length f_{L2}. Hence, the two lenses share a common focal plane. The plane of focus resides at another focal length f_{L2} away from the second lens. In total, the plane of focus is at a distance of $2 \cdot f_{L1} + 2 \cdot f_{L2}$ from the microlens array, i.e., to the image plane, which is why the setup is called a $4f$-light field camera. By means of this optical configuration, all light rays that are deflected by the same angle α in the measurement field will reach the image sensor with the same spatial offset δ_α with respect to the projected center of the corresponding microlens. Therefore, any displacement δ on the sensor can be unambiguously mapped to a deflection angle. By illuminating a transparent test object with collimated light from one side and by observing its other side with such a $4f$-light field camera, deflection maps as introduced in Section 4.1 can be acquired.

The advantage of the $4f$-light field camera of providing a mapping of angular to spatial deflection information that is independent of the spatial position of the deflection event in the measurement field comes at the disadvantage that the lateral size of the measurement field is reduced to the diameter of the second lens, similar to telecentric lenses.

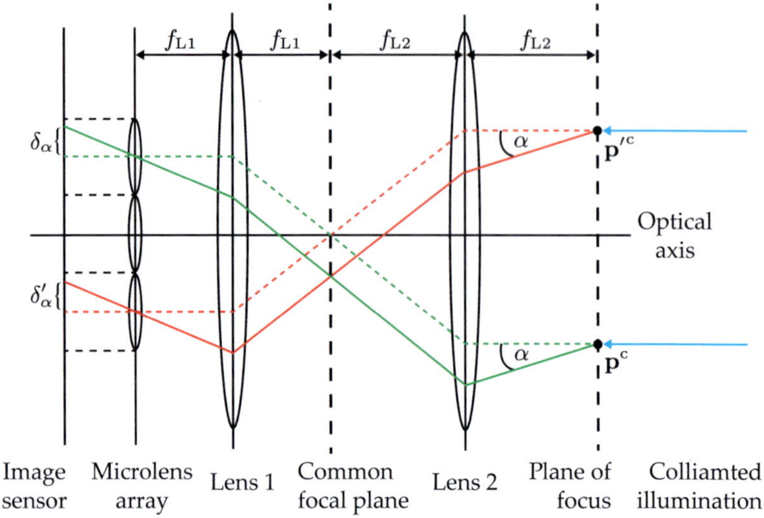

Figure 4.8: Principle of the optical setup of the $4f$-light field camera for the acquisition of deflection maps. Similar to a standard light field camera, the image sensor is placed in the focal plane of a microlens array. The microlenses reside in the focal plane of the first lens (with focal length f_{L1}), which shares its focal plane on the right-hand side with a second lens (with focal length f_{L2}). The plane of focus is at a distance of f_{L2} from the second lens. If two incident rays (red and green) of a collimated illumination are deflected by an angle α at the plane of focus, they reach the image sensor with the same displacement of $\delta_\alpha = \delta'_\alpha$ with respect to the projected center of the corresponding microlenses.

The $4f$-light field camera has been realized as a plugin for the Mitsuba rendering framework that is described in Chapter 5. With this plugin, experiments can be performed by simulating deflection maps of transparent objects and by processing them with the processing methods described in Section 6. The results of the experiments will be presented in Section 6.1.3.1. Figure 4.9 shows a simulated example of a light field image of a transparent test object obtained with a $4f$-light field camera.

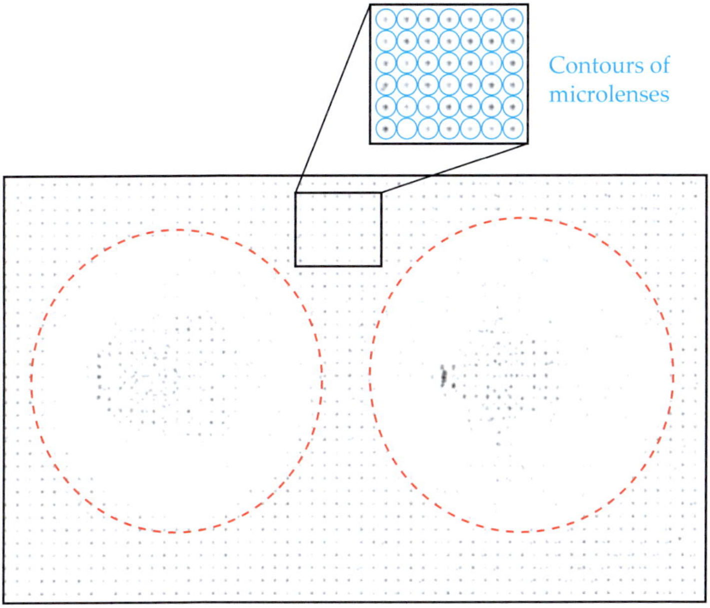

Figure 4.9: Example section of a light field image. The main image is an inverted (bright pixels correspond to low light intensity on the sensor and dark pixels correspond to high light intensity) gray value inspection image of a double-convex lens obtained with a simulated $4f$-light field camera under collimated illumination. The regular dot-like pattern (also magnified in the upper rectangle) is caused by the fact that only the central pixels under every microlens have received incident light, since the collimated illuminating light rays have not been deflected. The two dashed red circles indicate defective regions of the test object affected by enclosed air bubbles.

4.3 Schlieren light field imaging

The concept of schlieren imaging, which was introduced by August Toepler in 1864, can also be employed to capture the direction of light rays transmitted through transparent material [Set01]. The original schlieren setup is described in Section 4.3.1 and Section 4.3.2 explains a modified implementation of the concept, the schlieren deflectometer.

4.3.1 Original schlieren imaging

Figure 4.10 illustrates the concept of the original schlieren method. A collimated light source illuminates the test object from one side. The object's other side is focused on and observed by a specially modified camera based on the thin lens model described in Section 4.2. In the schlieren setup, a so-called *schlieren stop* (often also called a *knife edge*) is positioned at the image side focal plane of the imaging lens. As mentioned before (see Eq. (4.6) on p. 64), a lens transforms the angle by which a light ray gets deflected with respect to the optical axis into a spatial displacement on the image side focal plane of the imaging lens. The schlieren stop acts as a filter that lets only those rays pass whose direction of propagation in the measurement field corresponds to displacements on the focal plane not covered by the schlieren stop. By this means, images can be acquired that are formed only by rays with defined propagation directions. Hence, the schlieren setup can also be used to obtain deflection maps for transparent objects. By placing a transparent test object at the plane of focus inside the measurement field and by capturing images for different configurations of the schlieren stop, the object's deflection map can successively be acquired. A computationally configurable schlieren stop can be realized using a spatial light modulator,

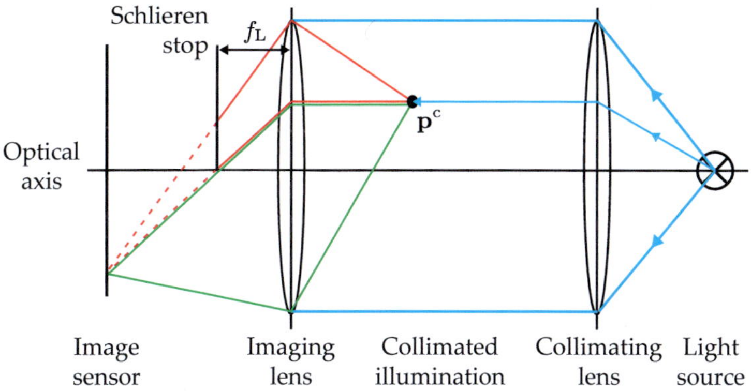

Figure 4.10: Principle of the optical setup of the original schlieren imaging. The measurement field is illuminated with a collimated light beam, which can be realized, e.g., by placing a point light source in the focal plane of a lens. Deflections inside the measurement field, e.g., at the object point $\mathbf{p}^c$, cause the affected rays to cross the focal plane of the observing lens with a spatial displacement corresponding to the deflection angle. Depending on the extent of the deflection, the rays can get blocked by the schlieren stop (red light bundle). By this means, the image observed by the sensor is formed only by rays with deflection angles not blocked by the schlieren stop (green light bundle).

e.g., with a digital micromirror device (DMD) or with a transmissive (LCD) or reflective (LCoS[3]) liquid crystal display [Mey14]. However, depending on the size of the controllable elements of the spatial light modulator, diffraction effects can lead to severe artifacts, which have to be coped with.

Such schlieren setups are mainly used to visualize differences in refractive index distributions, e.g., in gas flows or fluids. Figure 4.11 shows examples of images visualizing the gas flow of a cigarette lighter.

[3] Liquid crystal on silicon.

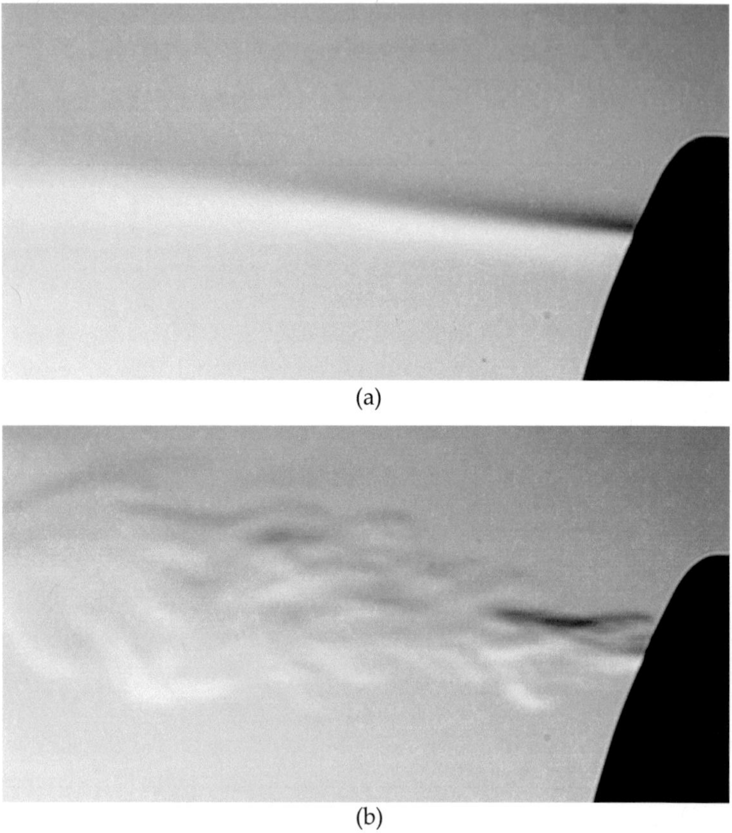

(a)

(b)

Figure 4.11: Examples of images visualizing the gas flow out of a cigarette lighter (black object on the right) obtained with an original schlieren imaging setup: (a) laminar gas flow; (b) turbulent gas flow.

4.3.2 Schlieren deflectometer

Another concept to obtain deflection maps of transparent test objects based on the principle of schlieren imaging is the so-called *schlieren deflectometer* [Sud13; Sud15]. Figure 4.12 shows the corresponding optical setup. For the schlieren deflectometer, the roles of the illumination component and the imaging component are swapped from those of the original schlieren setup. Here, the test object is observed with a telecentric camera which lets only those rays contribute to the acquired image that propagate parallel to the optical axis inside the measurement field. Furthermore, the test object is not illuminated with a single collimated light beam oriented parallel to the optical axis, but with several collimated light beams tilted with respect to the optical axis in a time-sequential manner. This illumination is realized by placing a two-dimensional programmable light source, e.g., a computer display, in the focal plane of a lens. A collimated light beam tilted by an angle of α with respect to the optical axis can be emitted by turning on the pixel of the display having a distance of $\delta_\alpha = f_\mathrm{L} \cdot \tan(\alpha)$ from the optical axis, with f_L denoting the focal length of the collimating lens. If a ray of the illuminating collimated light bundle is deflected so that it propagates parallel to the optical axis, it passes the camera's telecentric stop, which leads to a high intensity on the image sensor. Any pixel of the acquired image having a high gray value indicates that in the plane of focus in the measurement field an illuminating light bundle has been transformed into a light bundle propagating parallel to the optical axis. By acquiring a series of images for multiple tilted collimated light beams, i.e., for different pixels of the light source with varying distances δ_α from the optical axis turned on sequentially, a deflection map of a transparent test object can be obtained. Each single image corresponds to a two-dimensional (j,k)-slice through the deflection map $a(m,n,j,k)$. In contrast to the original definition of the deflection maps given in Section 4.1 assuming the test object to be illuminated with a collimated light beam parallel to the optical axis and to be observed by a sensor sensitive to the direction of propagation, the schlieren deflectometer illuminates the test object with collimated light beams of multiple directions and captures only light bundles parallel to the optical axis. In other words, the roles of the illuminating and capturing components are interchanged, leading to an inversion of the direction of the light transport according to the Helmholtz reciprocity principle [Hap93].

If only a single pixel is turned on simultaneously, a technical realization with a display can result in a weak illumination and hence in a reduced signal-to-noise ratio. This can be improved by multiplexing multiple pixels according to certain patterns for the acquisition of each image and by computationally demultiplexing them afterwards to obtain the images corresponding to each single illumination pixel.

Figure 4.12: Principle of the optical setup of a schlieren deflectometer. A 2D programmable area light source is placed in the focal plane of a lens to realize a tiltable collimated illumination. A telecentric camera system observes the measurement field. If an illuminating light ray is deflected, e.g., at point $\mathbf{p}^c$, so that it propagates parallel to the optical axis, it passes the camera's telecentric stop and contributes to the acquired image.

For example, so-called *Hadamard codes* provide suitable multiplexing schemes, since they rely on binary patterns where the pixels are either completely turned on or off [Sch07]. Figure 4.13 shows images of a deflection map of a double-convex lens acquired with a schlieren deflectometer.

4.4 Light field imaging based on a laser scanner

The described camera-based systems for acquiring deflection maps all share the disadvantages of a comparatively small measurement field and a reduced depth of field. Laser scanner systems can mitigate these drawbacks by performing a time-sequential sampling of the measurement field with collimated laser beams. Section 4.4.1 introduces a laser scanner system suitable for capturing deflection maps of transparent objects and Section 4.4.2 shows how the optical setup can be extended by means of a position sensitive detector.

4.4.1 Laser deflection scanner

Figure 4.14 shows the principle of the optical setup of a laser deflection scanner. The system consists of two main components: an emitter and a receiver. The emitter illuminates the measurement field with parallel laser beams in a time-sequential manner, i.e., one laser beam at a time, realizing a successive sampling of the measurement field. This can be optically realized, e.g., by a laser beam source that shines on a rotating polygonal mirror wheel which is positioned at the focal point of a lens or a parabolic mirror.

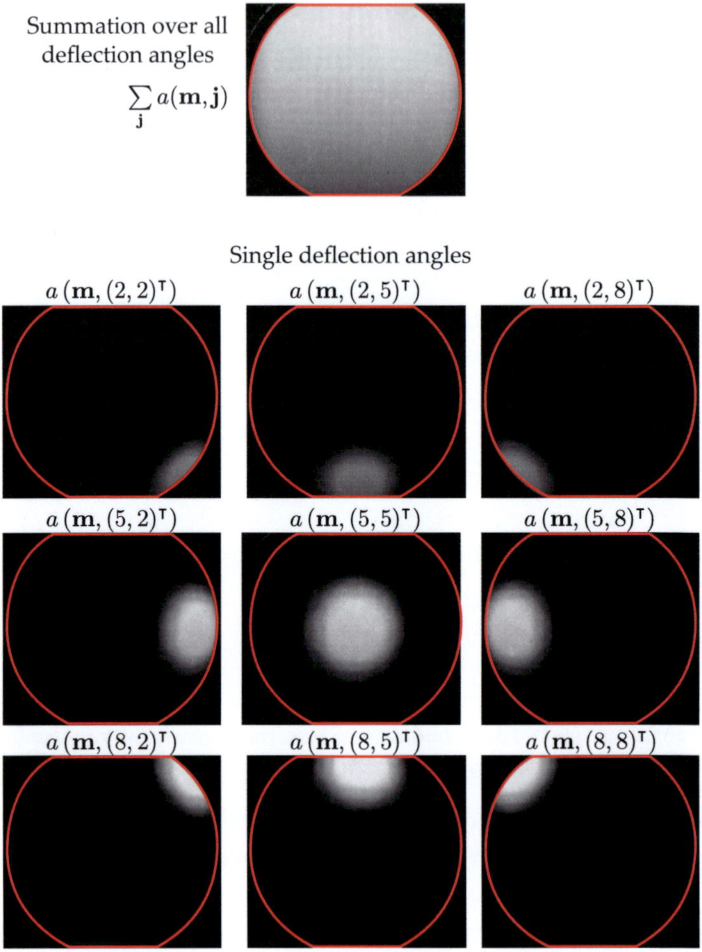

Figure 4.13: Examples of images of the acquisition of a deflection map of a double-convex lens by means of a schlieren deflectometer. The top image shows the summation of the images over all illumination angles used. The other nine images show sections of the acquired deflection map for selected illumination angles. The angular coordinates $(5,5)^\mathsf{T}$ correspond to undeflected rays. The red marking highlights the borders of the test object.

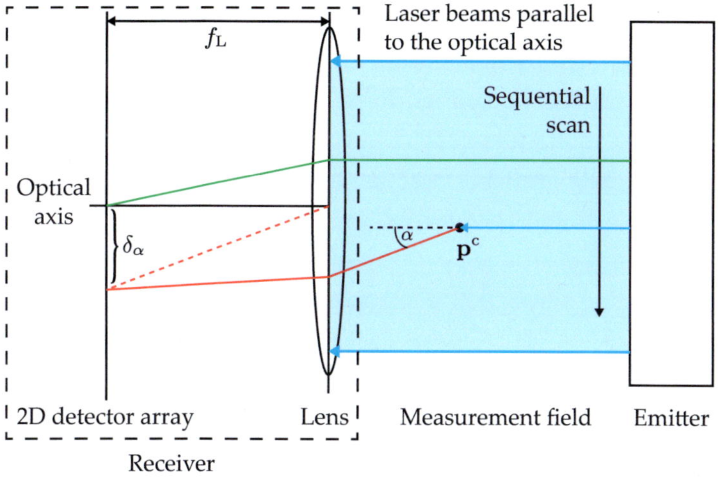

Figure 4.14: Principle of the optical setup of a laser deflection scanner. An emitter illuminates the measurement field with laser beams parallel to the optical axis in a time-sequential manner. Laser beams reaching the receiver with a direction of incidence parallel to the optical axis will hit the center of the two-dimensional detector array (green ray). Beams that are deflected by an angle α in the measurement field, e.g., at point $\mathbf{p}^c$, hit the detector array with a spatial displacement of $\delta_\alpha = f_L \cdot \tan(\alpha)$ with respect to the center of the detector array (red ray). In this way, two-dimensional deflection distributions are captured for each position of the laser beam.

The receiving component consists of a lens or a parabolic mirror having a two-dimensional detector array placed in its focal plane. Again, the principle of transforming angular information about the direction of propagation of the light into a spatial displacement is employed here. If a laser beam is deflected in the measurement field by an angle α with respect to the optical axis, it will hit the two-dimensional detector array at a distance of $\delta_\alpha = f_L \cdot \tan(\alpha)$ from the optical axis. Light bundles propagating parallel to the optical axis through the measurement field, i.e., without being deflected, are directed to the center of the detector array.

The emitter and the receiver are synchronized, so that the electrical signals generated by the detector array can be read for every single laser beam shifted in parallel. By this means, a complete deflection map of a transparent test object can be captured. Since the detectors of commercially available laser scanners can achieve readout rates in the MHz range, short acquisition periods can be realized. Figure 4.15 shows images of a deflection map of a washing machine door glass acquired using a laser deflection scanner.

4.4.2 Combination with a position sensitive detector

The sensor component of the introduced laser deflection scanner can be extended or even replaced by a so-called *position sensitive detector* (PSD). Figure 4.16 shows the corresponding optical setup.

A PSD has a two-dimensional photosensitive sensor area that allows measuring the 2D-position of an incident light spot. If the incident light is not shaped like a spot, the PSD yields the center of gravity of the incident light pattern, i.e., a sum of the single positions of the pattern weighted with the respective intensity. Via a beam splitter, the intensity distribution incident on the 2D detector array can be mirrored to the PSD-sensor. By this means, the PSD-signal yields the mean deflection direction $\mathbf{w}(\mathbf{m}) = (w_x(\mathbf{m}), w_y(\mathbf{m}))^\mathsf{T}$ of the current measurement point in x- and y-direction.

The information obtained with a PSD (the center of gravity of the incident light pattern) is less than with the general laser deflection sensor (the full deflection map). However, for the detection of some defects, the PSD coordinates can be sufficient, as will be shown in Section 6.2. This can be advantageous, since it requires much less time to transfer and process the PSD coordinates than it does with the intensities recorded by the detector array of the standard laser deflection scanner.

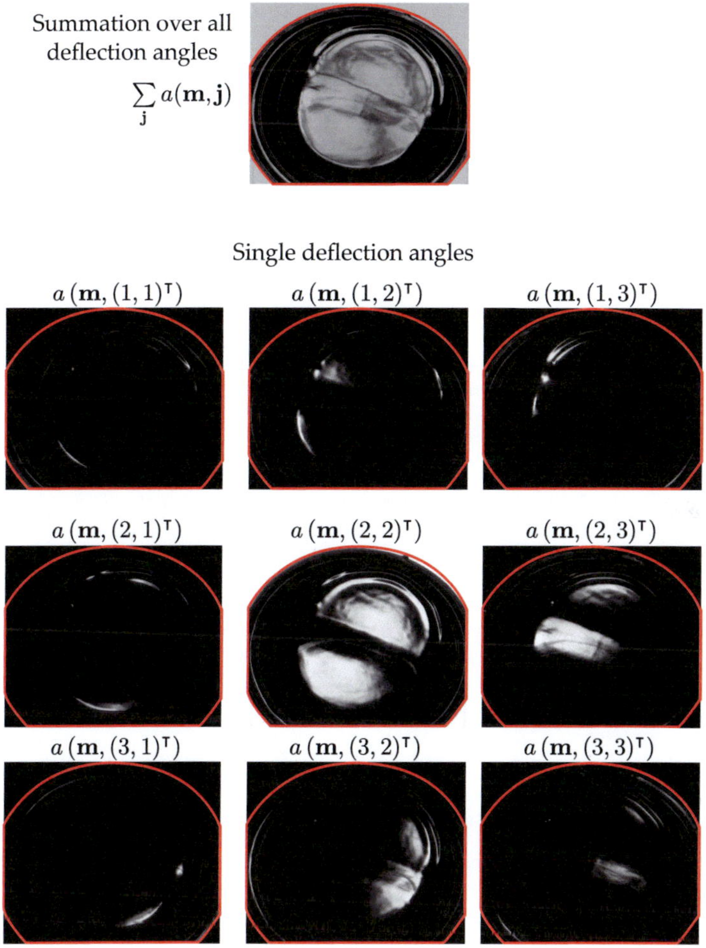

Figure 4.15: Example images of the acquisition of a deflection map of a washing machine door glass by means of a laser deflection scanner. The top image shows the summation of the images of all captured deflection angles. The other nine images show sections of the acquired deflection map for all 3×3 deflection angles supported by the laser scanner. The angular coordinates $(2,2)^{\mathsf{T}}$ correspond to undeflected rays. The red marking highlights the borders of the test object.

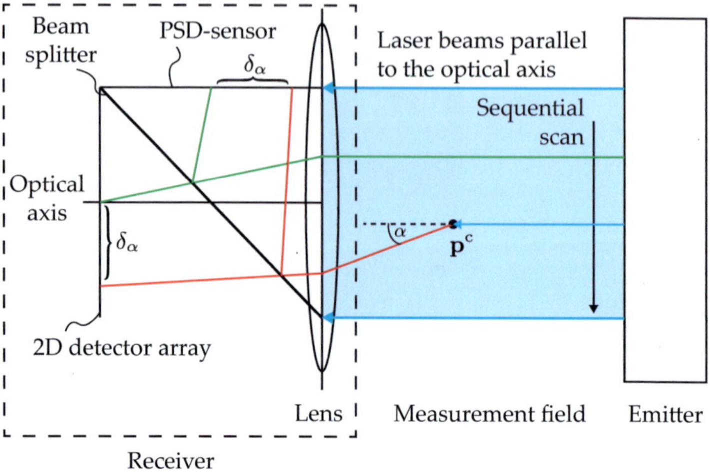

Figure 4.16: Extension of the laser deflection scanner by a position sensitive detector (PSD). The original optical setup (cf. Fig. 4.14) is extended by a beam splitter and a PSD inside the receiver component. Incident light rays hit the 2D detector array and the PSD with the same spatial displacements that depend on the corresponding deflection angle in the measurement field.

4.5 Light transport matrices

Light transport matrices are another concept for describing the light transport of a scene, they can also contain information about the light's direction of propagation. For a scene with Q light sources and S sensor elements, the light transport matrix $\mathbf{T} \in \mathbb{R}^{S \times Q}$ holds the contributions of every light source to every pixel. The element $T_{m,n}$ represents the contribution of light source n to detector element m.

Light transport matrices can be acquired in various ways, as was described in Section 2.1.6. Figure 4.17 shows an optical setup to obtain light transport matrices, suitable for the visual inspection of transparent objects. A spatially programmable two-dimensional light source, e.g., a display device, illuminates the test object from

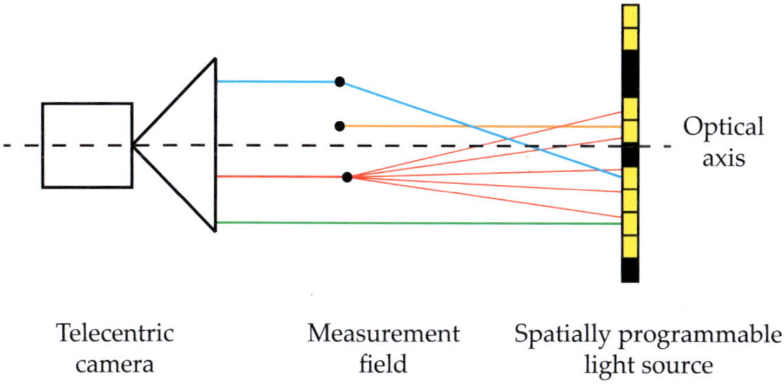

Figure 4.17: Principle of the optical setup for the acquisition of light transport matrices for the visual inspection of transparent objects. A telecentric camera is focused on the plane of interest in the measurement field. A spatially programmable light source illuminates the measurement field from the other side. Depending on the objects inside the measurement field, the signal of a pixel inside the camera can be influenced by illuminating light rays that either propagate straightly towards the camera (green ray), by light rays whose direction is changed due to refraction or reflection (blue ray) or even by multiple illuminating light rays that are scattered towards the camera, e.g., due to a scattering material defect (red rays). Indeed, light rays can also be absorbed or reflected outside of the optical system so that they do not reach the camera at all (orange ray, absorbed).

one side. The object's other side is observed by a telecentric camera system. A naive way to acquire the scene's light transport matrix is to capture a series of images while sequentially turning on every single pixel of the display. Every image of the series represents a column of the matrix $\mathbf{T}$. However, this acquisition strategy has a poor light efficiency, as mentioned before (see Section 4.3.2). Again, employing a series of multiplexed illumination patterns and performing a computational demultiplexing can lead to an improved light efficiency. By means of the methods described in Section 6.4, the acquired light transport matrix can be processed in order to reveal scattering material defects inside the test object.

If no test object is placed inside the measurement field, the rays of sight of the camera will propagate straight to the light source. In this case, depending on the relative sizes of the camera's detector elements, the pixels of the spatially programmable light source, and the magnification of the camera's lens system, only one or some adjacent pixels of the light source will contribute to a single detector element of the camera. Hence, the rows of the light transport matrix will have only a few spatially adjacent non-zero elements. This is also the case for completely defect-free transparent test objects, since they only change the direction of propagation of the transmitted light rays, but do not cause any scattering that would lead to more light source pixels contributing to single detector elements.

When considering state-of-the-art cameras with up to 20 million detectors and state-of-the-art spatially programmable light sources with over 1 million pixels (e.g., a computer display), the concept of light transport matrices becomes impractical due to the issues of acquisition time and computer memory. The light transport matrix for these parameters would have $20 \cdot 10^6 \cdot 1 \cdot 10^6 = 20 \cdot 10^{12}$ elements, requiring an acquisition time of over nine hours (assuming a camera frame rate of 30 frames per second) and computer memory of 20 terabytes (assuming 8-bit quantization). However, the light transport matrices of scenes such as those shown in Fig. 4.17 with transparent test objects are usually sparse, i.e., most of the elements of $\mathbf{T}$ are equal to zero. Therefore, $\mathbf{T}$ can be reconstructed out of a markedly reduced number of acquired images by means of methods following the approach of compressive sensing [Pee09]. Depending on the visual inspection application, illumination patterns with a far lower resolution can be sufficient, which enables a brute force acquisition in a reasonable time and with limited computer memory. Section 6.4.2 describes conducted experiments where simulated light transport matrices have been processed in order to visualize material defects inside transparent test objects.

4.6 Summary

This section has introduced various optical setups for the acquisition of light field images, as well as those for the acquisition of light deflection maps, assuming the investigated test object to be illuminated with collimated light. Based on the thin lens camera model, the concept of imaging with a lens and of sensing deflection angles in the focal plane of the lens has been explained. The latter principle is used by most optical setups for capturing light fields. As an extension to common light field cameras, a $4f$-light field camera setup has been described, which enables a sensitivity to deflection angles that is laterally constant over the whole measurement field.

Furthermore, the principle of schlieren imaging has been explained. Related optical setups rely on applying spatial filters to the focal plane of a lens in order to obtain images formed only by light bundles with selected propagation directions. Based on the schlieren principle, the schlieren deflectometer has been described. In this setup, a deflection map of the test object is obtained by observing it with a telecentric camera and by time-sequentially illuminating it with collimated light beams that are tilted by different angles with respect to the optical axis.

The introduced deflection laser scanners illuminate the test object with parallel laser beams in a time-sequential manner and capture the direction of the transmitted rays by means of a sensor placed in the focal plane of a parabolic mirror. As the sensing component captures two-dimensional angular information and the emitted parallel laser performs a spatial sampling of the test object, four-dimensional deflection maps are obtained. In addition, the basic optical setup of these deflection laser scanners can be extended by a position sensitive detector. For these sensors, only the coordinates of the mean deflection direction have to be transferred and processed, which increases the potential inspection speed but simultaneously decreases the available amount of information. However, the PSD signals are sufficient for some inspection tasks.

As a further approach to obtain light field related information about transparent test objects, the concept and acquisition of light transport matrices have been introduced. Light transport matrices store the contributions of an inspection scene's light sources to every pixel of the observing camera. If the test object is illuminated by a spatially programmable light source and the transmitted light is observed by a telecentric camera, the resulting light transport matrix can be processed to visualize material defects (see Section 6.4).

5 Simulation framework

As mentioned in Chapter 1, designing a conventional machine vision setup suitable for automated visual inspection can be a tedious task that involves many iterations until the desired performance is reached. To reduce the costs of continuously manually adjusting system parameters (e.g., the sensor or illumination placement), simulation software can be employed to computationally synthesize camera images or sensor signals of a digitally represented machine vision system. The simulation environment is required to produce as physically realistic images as possible. The *Mitsuba renderer* [Jak10] is a photorealistic renderer in the style of the famous *Physically Based Renderer (PBRT)* [Pha16], which can be extended by customized add ons, so-called *plugins*. This allows novel sensor concepts or light sources to be added to Mitsuba.

Many of the concepts introduced in this thesis have been simulated using Mitsuba before building a physical prototype to validate the ideas. Several different plugins have been developed to enhance the set of sensors, light sources and simulation capabilities of Mitsuba. In Section 5.1, this chapter provides a short introduction to the Mitsuba renderer by outlining its fundamental structure. Section 5.2 describes the developed plugins and Section 5.3 closes the chapter with a summary.

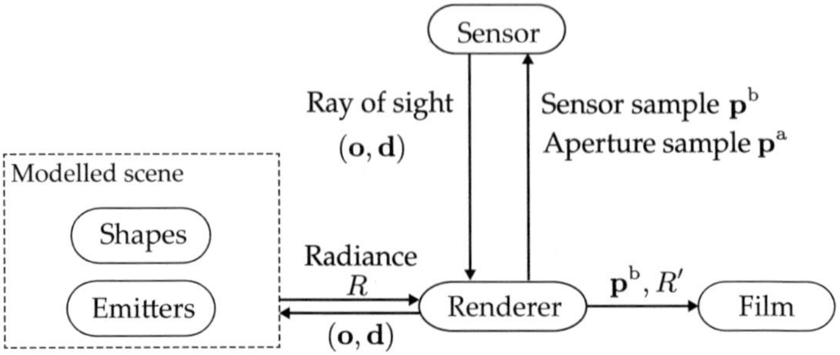

Figure 5.1: Illustration of the fundamental framework components of Mitsuba.

5.1 Rendering framework

The Mitsuba renderer consists of five main components whose interaction is illustrated in Figure 5.1. Every scene that should be rendered with Mitsuba consists of at least a sensor component, such as a camera, and a light emitter. Further objects can be composed out of so-called shapes or they can be added by means of their CAD model[1]. To synthesize the sought camera image, a classical ray tracing algorithm is employed: The renderer component generates two-dimensional continuous sensor samples $\mathbf{p}^{\mathrm{b}}$ according to a certain sampling strategy and, if requested by the sensor plugin, additional aperture samples $\mathbf{p}^{\mathrm{a}}$ and passes them to the sensor plugin. For example, an aperture sample is needed by plugins modeling conventional cameras (e.g., the thin lens camera model, cf. p. 61), where every pixel integrates incident light rays coming from multiple directions. The sensor calculates the ray of sight corresponding to the received spatial sample $\mathbf{p}^{\mathrm{b}}$ and, if applicable, the aperture sample $\mathbf{p}^{\mathrm{a}}$ according to its sensor model. Rays of sight are defined as tuples $(\mathbf{o}, \mathbf{d})$ of a

[1] Here, the term CAD (computer aided design) model refers to a three-dimensional triangle mesh representing the object of interest.

point of origin $\mathbf{o}$ and the direction of propagation $\mathbf{d}$. The ray of sight is passed back to the renderer, which starts tracing the ray through the modeled scene. The ray might hit the surface of an object, get absorbed, reflected or refracted and might receive new directions of propagation. The tracing is continued until no further intersection happens (i.e., the ray directs into empty space), a fixed number of intersections is exceeded or a light source has been reached. In the latter case, the reached light source component is queried for its radiance R with respect to the current ray of sight.[2] The renderer propagates the received radiance R back along the whole optical path of the traced ray by taking all optical effects that occur along the way and the reflectance characteristics of all involved surfaces into account. The resulting radiance R' and the current sensor sample $\mathbf{p}^{\mathrm{b}}$ are passed to the film component which successively aggregates the received radiances and it finally creates the output image.

5.2 Introduced rendering framework components

The Mitsuba framework already contains plugins for modeling several sensors, light sources, and so on. Since the main target application of Mitsuba is to create photorealistic images of natural scenes, it is missing some components that are relevant for simulating machine vision setups. The following sections describe the plugins that are contributed to the Mitsuba framework by this thesis and that allow to simulate many machine vision setups, even such employing light field methods.

[2] For the scope of this thesis, the light's wavelength λ is neglected in the rendering process. The Mitsuba framework could also handle multiple radiances corresponding to different spectral bands.

5.2.1 Sensor plugins

A Mitsuba sensor plugin has to provide a function that calculates the parameters of the ray of sight $(\mathbf{o}, \mathbf{d})$ corresponding to a pixel sample $\mathbf{p}^{\mathrm{b}} = (p_x^{\mathrm{b}}, p_y^{\mathrm{b}})^\intercal$ and—if required—an aperture sample $\mathbf{p}^{\mathrm{a}} = (p_x^{\mathrm{a}}, p_y^{\mathrm{a}})^\intercal$. In the following, the introduced sensor plugins are described and the steps needed for determining the rays of sight corresponding to the individual sensor models are presented.

5.2.1.1 $4f$-light field camera

The optical setup of the $4f$-light field camera has been introduced in Section 4.2.2. The sensor is placed in the focal plane of an array of rectangular microlenses which is located in the focal plane of a first lens with focal length f_{L1}. The focal plane of the other side of this lens coincides with the focal plane of a second lens with focal length f_{L2}. The resulting plane of focus in the object space is located in the object side focal plane of the second lens.

The calculation of the ray of sight corresponding to a sensor sample $\mathbf{p}^{\mathrm{b}}$ and an aperture sample $\mathbf{p}^{\mathrm{a}}$ involves several steps for this sensor model. Figure 5.2 illustrates the calculation of the ray of sight. If not stated differently, all involved points are two-dimensional and their position along the optical axis will be specified accordingly. As a first step, the intersection $\mathbf{p}^{\mathrm{f}}$ of the ray of sight and all rays of the collimated ray bundle captured by the microlens corresponding to $\mathbf{p}^{\mathrm{b}}$ with the common focal plane of lens 1 and lens 2 is determined. This intersection results from calculating the slope of the principal ray (dashed green line) of the microlens and by following a ray with that slope for a distance of f_{L1} beginning from the center of lens 1:

$$\mathbf{p}^{\mathrm{f}} = \frac{f_{\mathrm{L1}}}{f_{\mathrm{ML}}} \cdot \left(\mathbf{c}^{\mathrm{ML}} - \mathbf{p}^{\mathrm{b}}\right) , \tag{5.1}$$

with $\mathbf{c}^{\mathrm{ML}}$ denoting the center of the corresponding microlens on the plane of the microlens array. The intersection $\mathbf{p}^{\mathrm{L1}}$ of the optical path of the ray of sight with lens 1 can now be obtained via:

$$\mathbf{p}^{\mathrm{L1}} = \mathbf{p}^{\mathrm{a}} + \mathbf{p}^{\mathrm{f}}\,. \tag{5.2}$$

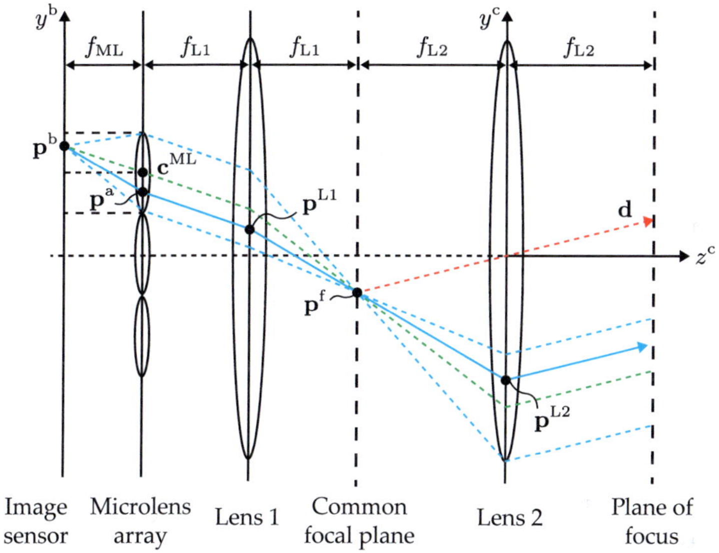

Figure 5.2: Calculation of the ray of sight for the $4f$-light field camera sensor plugin. The solid blue line shows the optical path of the ray of sight corresponding to the sensor sample $\mathbf{p}^{\mathrm{b}}$ and the aperture sample $\mathbf{p}^{\mathrm{a}}$. The dashed blue lines denote the whole light bundle that is captured by the microlens and the dashed green line represents the principal ray of the microlens. The dashed red arrow illustrates how the direction $\mathbf{d}$ of the ray of sight is obtained.

Calculating the intersection of a ray which starts from $\mathbf{p}^{\mathrm{L1}}$ and runs through $\mathbf{p}^{\mathrm{f}}$ with lens 2 yields $\mathbf{p}^{\mathrm{L2}}$, that is, the sought intersection of the optical path of the ray of sight with lens 2:

$$\mathbf{p}^{\mathrm{L2}} = \mathbf{p}^{\mathrm{L1}} + \left(\mathbf{p}^{\mathrm{f}} - \mathbf{p}^{\mathrm{L1}}\right) \frac{f_{\mathrm{L1}} + f_{\mathrm{L2}}}{f_{\mathrm{L1}}} . \tag{5.3}$$

This equation can be simplified by substituting the second occurrence of $\mathbf{p}^{\mathrm{L1}}$ with Eq. (5.2):

$$\mathbf{p}^{\mathrm{L2}} = \mathbf{p}^{\mathrm{L1}} - \mathbf{p}^{\mathrm{a}} \frac{f_{\mathrm{L1}} + f_{\mathrm{L2}}}{f_{\mathrm{L1}}} . \tag{5.4}$$

The origin $\mathbf{o}$ of the sought ray of sight is given by $\mathbf{p}^{\mathrm{L2}}$:

$$\mathbf{o} = \begin{pmatrix} o_x \\ o_y \\ o_z \end{pmatrix} = \begin{pmatrix} p_x^{\mathrm{L2}} \\ p_y^{\mathrm{L2}} \\ 0 \end{pmatrix} . \tag{5.5}$$

When calculating $\mathbf{p}^{\mathrm{L1}}$ and $\mathbf{p}^{\mathrm{L2}}$, it has to be evaluated whether or not the calculated points are still on the active area of the lenses. If this is not the case, then the calculation of the ray of sight is aborted and a spectrum with all elements set to zero is passed back to the renderer. By this means, the effect of vignetting is taken into account.

Since all rays of the whole bundle propagate through $\mathbf{p}^{\mathrm{f}}$, which resembles a point source placed in the focal plane of lens 2, these rays propagate parallel after being transmitted through lens 2. The sought direction $\mathbf{d}$ is given by a ray originating from $\mathbf{p}^{\mathrm{f}}$ and running through the center of lens 2, that is, by the principal ray of lens 2 with respect to $\mathbf{p}^{\mathrm{f}}$:

$$\mathbf{d} = \begin{pmatrix} d_x \\ d_y \\ d_z \end{pmatrix} = \begin{pmatrix} -p_x^{\mathrm{f}} \\ -p_y^{\mathrm{f}} \\ f_{\mathrm{L2}} \end{pmatrix} . \tag{5.6}$$

The $4f$-light field camera plugin is parameterized as follows. The number of microlenses $M_{\mathrm{ML}}, N_{\mathrm{ML}}$ in horizontal and vertical direction determine the spatial resolution. Specifying the angular resolution $A_{\mathrm{h}}, A_{\mathrm{v}}$ and the pitches of the microlenses $s_{\mathrm{h}}, s_{\mathrm{v}}$ in horizontal and vertical directions also defines the size of the pixels and the overall size of the image sensor. Additionally, the three focal lengths $f_{\mathrm{ML}}, f_{\mathrm{L1}}, f_{\mathrm{L2}}$ and the diameters of lens 1 and lens 2 have to be provided.

5.2.1.2 Light field sensor

The light field sensor plugin represents an abstract plugin that can be employed to capture the light field at a certain rectangular region in space. The plugin's concept is similar to that of a *Shack–Hartmann* wavefront sensor [Pla01]. The light field sensor is based on an array of rectangular microlenses, which is placed in front of the image sensor with a distance of its focal length f_{ML}. The calculation of the ray of sight $(\mathbf{o}, \mathbf{d})$ for a sensor sample $\mathbf{p}^{\mathrm{b}}$ and an aperture sample $\mathbf{p}^{\mathrm{a}}$ is illustrated by Figure 5.3. Again, all points are two-dimensional if not stated differently and their location along the optical axis is specified accordingly. The ray of sight's origin $\mathbf{o}$ can be directly obtained via the aperture sample:

$$\mathbf{o} = \begin{pmatrix} o_x \\ o_y \\ o_z \end{pmatrix} = \begin{pmatrix} p_x^{\mathrm{a}} \\ p_y^{\mathrm{a}} \\ 0 \end{pmatrix} . \tag{5.7}$$

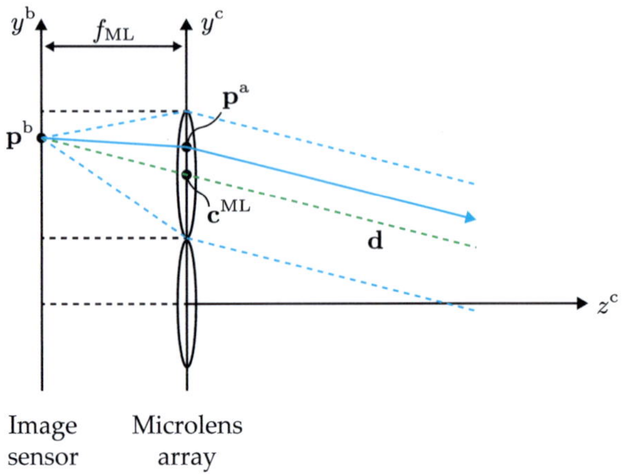

Figure 5.3: Illustration of the calculation of a ray of sight for the light field sensor plugin. The blue arrow denotes the ray of sight corresponding to the sensor sample $\mathbf{p}^{\mathrm{b}}$ and the aperture sample $\mathbf{p}^{\mathrm{a}}$. The point $\mathbf{c}^{\mathrm{ML}}$ represents the center of the corresponding microlens. The dashed blue lines show the whole light bundle that is incident on $\mathbf{p}^{\mathrm{b}}$ and the dashed green line represents the principal ray which also defines the direction $\mathbf{d}$ of the sought ray of sight.

The direction $\mathbf{d}$ is given by the principal ray of the corresponding microlens, which runs through $\mathbf{p}^{\mathrm{b}}$ and the center $\mathbf{c}^{\mathrm{ML}}$ of the corresponding microlens. The vector $(\mathbf{c}^{\mathrm{ML}} - \mathbf{p}^{\mathrm{b}})$ and the focal length f_{ML} determine the slope of this ray. Hence, the sought direction is calculated via:

$$\mathbf{d} = \begin{pmatrix} d_x \\ d_y \\ d_z \end{pmatrix} = \begin{pmatrix} c_x^{\mathrm{ML}} - p_x^{\mathrm{b}} \\ c_y^{\mathrm{ML}} - p_y^{\mathrm{b}} \\ f_{\mathrm{ML}} \end{pmatrix} . \tag{5.8}$$

The captured four-dimensional light field is spatially multiplexed on the image sensor as illustrated by Figure 5.4.

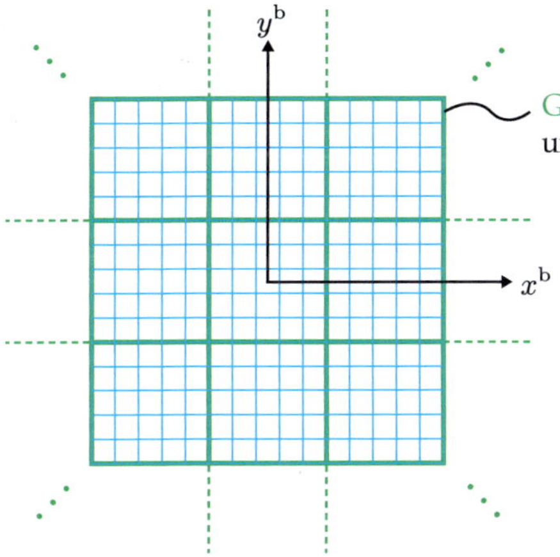

Figure 5.4: Spatial multiplexing by the light field sensor plugin. The blue rectangles represent single pixels of the image sensor. The green rectangles denote groups of pixels belonging to a common microlens. Inside a group of pixels, every pixel corresponds to a certain direction of propagation of captured light bundles.

All adjacent sensor pixels underneath one microlens correspond to light bundles with the same origin (i.e., with the same spatial component), but with a different direction of propagation (i.e., a different angular component).

The light field sensor plugin is parameterized as follows. Its spatial resolution $M_{\mathrm{ML}}, N_{\mathrm{ML}}$ in horizontal and vertical direction determines the number of microlenses in the two dimensions. The angular resolution $A_{\mathrm{h}}, A_{\mathrm{v}}$ denotes the number of pixels under one microlens in horizontal and vertical direction. Together, the spatial resolution and the angular resolution determine the total resolution $M = M_{\mathrm{ML}} \cdot A_{\mathrm{h}}, N = N_{\mathrm{ML}} \cdot A_{\mathrm{v}}$ of the image sensor in horizontal and vertical direction. The pitches $s_{\mathrm{h}}, s_{\mathrm{v}}$ of the microlenses in horizontal and vertical direction also have to be specified and define the overall size of the image sensor.

5.2.2 Emitter plugins

In Mitsuba the emitter plugins have to allow for the following functionality: for a given intersection $\mathbf{p}^{\mathrm{i}}$ of a traced ray of sight with a light source and the corresponding direction of incidence $\mathbf{d}$, the emitter plugin has to provide a function that calculates the resulting spectrum of the emitted light. The following sections describe all introduced emitter plugins and show how the spectra are calculated for a given query from the renderer.

5.2.2.1 Parallel emitter

The parallel emitter plugin realizes a light source that emits light of a specified spectrum only into directions approximately parallel to the surface normal $\mathbf{n}$ at the queried intersection $\mathbf{p}^{\mathrm{i}}$. Since full parallelism would lead to many traced rays receiving no light from such an emitter, light is considered to be emitted inside a cone having a certain angle α with respect to the surface normal. Figure 5.5 illustrates the concept of the parallel emitter.

The plugin has to be attached to an arbitrary shape (i.e., a geometric object) that provides the surface information.

If the angle between the queried direction $\mathbf{d}$ and the surface normal $\mathbf{n}$ is greater than α, a spectrum with all elements set to zero is returned, otherwise, the specified spectrum is returned.

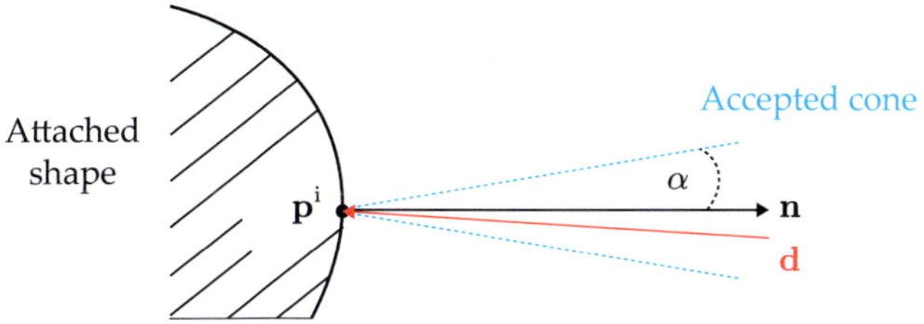

Figure 5.5: Concept of the parallel emitter plugin: the red arrow denotes the queried direction $\mathbf{d}$ and $\mathbf{n}$ represents the surface normal at the queried surface point $\mathbf{p}^{\mathrm{i}}$. Since the angle between $\mathbf{d}$ and $\mathbf{n}$ is less than the angle α of the accepted cone, light is emitted in the queried direction.

5.2.2.2 Spatially programmable area light source

A further plugin allows to simulate a spatially programmable area light source that consists of single pixels of the same size all emitting individual spectra. Figure 5.6 illustrates the concept of the plugin. The spectra can be conveniently defined by the user by providing an RGB-color image. Other file formats can be employed to support spectra of higher dimensions. The plugin has to be attached to a rectangular shape that determines the light source's overall size, position and orientation in the simulated scene. The rectangular shape and the number of pixels of the RGB-image in horizontal and vertical direction define the size of the emitter's pixels in the two dimensions. Similar to the parallel emitter plugin mentioned in the previous section, the pixels of this plugin emit light only inside a definable cone angle α.

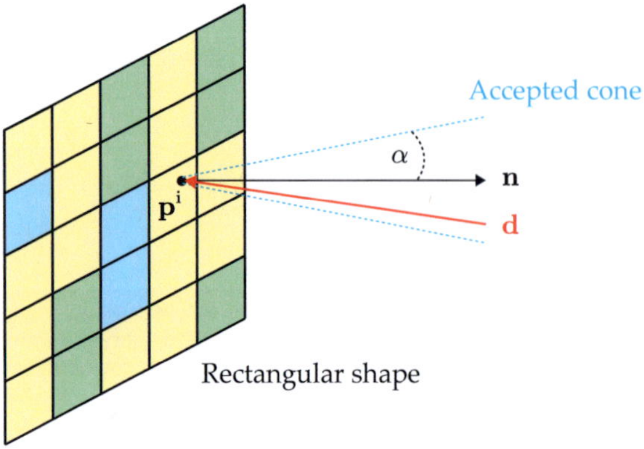

Figure 5.6: Concept of the spatially programmable area light source plugin: similar to the parallel emitter, every pixel of the simulated programmable area light source emits light inside a definable cone only. Since in the shown case the queried surface position lies inside a yellow pixel of the light source, the respective spectrum is returned to the renderer.

5.2.2.3 Light field emitter

The light field emitter plugin resembles the inverse of the light field sensor plugin described in Section 5.2.1.2. Its optical setup is similar to that of the light field generator introduced in Chapter 3 consisting of a two-dimensional display placed in the focal plane of an array of rectangular microlenses. The plugin requires an image file containing the four-dimensional light field to emit. The light field can be stored in a spatially multiplexed manner, as illustrated in Figure 5.4. Additionally, the focal length f_{ML} of the microlenses, the angular resolution in horizontal and vertical direction and a rectangular shape, to which the emitter is attached, have to be defined. The angular resolution, the number of pixels of the provided image file and the dimensions of the attached shape determine the emitter's spatial resolution and the size of the microlenses in horizontal and vertical direction.

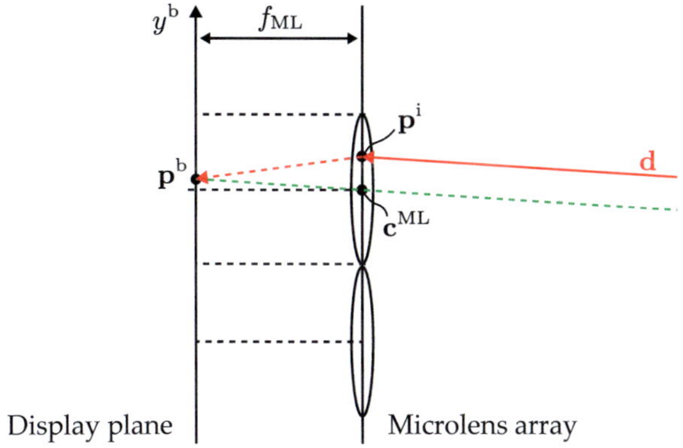

Figure 5.7: Concept of the light field emitter plugin. For a queried intersection $\mathbf{p}^i$ and direction $\mathbf{d}$ of incidence (solid red line) from the renderer, the corresponding pixel $\mathbf{p}^b$ of the display is determined. The dashed red line shows the optical path of the traced ray of sight after reaching the emitter. The green dashed line represents the principal ray of the microlens with respect to $\mathbf{p}^b$.

For a query from the renderer consisting of the position $\mathbf{p}^i$ of the intersection on the emitter and the direction $\mathbf{d}$ of the incident traced ray of sight, the plugin has to return the corresponding spectrum from the image file. Figure 5.7 illustrates the emitter's concept. The point $\mathbf{p}^i$ defines the spatial component, i.e., the hit microlens. To determine the pixel corresponding to the queried direction $\mathbf{d}$ inside the group of pixels under the determined microlens, the intersection $\mathbf{p}^b$ of a ray which starts from the center $\mathbf{c}^{ML}$ of the microlens in the direction $\mathbf{d}$ with the display plane is calculated. For the found intersection $\mathbf{p}^b$, which consists of continuous coordinates on the display plane, the RGB-values can be calculated, such as by means of a bilinear interpolation of the RGB-values of adjacent pixels of the provided image file [Bey15].

The introduced plugin does not take crosstalk into account, that is, if the angle between a queried direction and the surface of the emitter is small enough that the pixels under an adjacent microlens would be reached instead of those of the actually hit microlens, a spectrum with all elements set to zero is returned.

5.2.3 Simulation of light transport matrices

In Section 4.5, an optical setup for acquiring light transport matrices of transparent objects has been presented and Section 6.4 will show suitable methods for processing these matrices to detect scattering material defects. This section introduces an extension to the Mitsuba framework that allows to conveniently simulate light transport matrices for virtual scenes containing a spatially programmable light source. This extension has been employed to simulate the light transport matrices in the experiments shown in Section 6.4.2.

Figure 5.8 illustrates the modifications applied to the Mitsuba framework (the added components are marked in blue). Whenever a traced ray of sight is incident to a light source consisting of multiple pixels, the involved light source plugin may also report a discrete coordinate $\mathbf{u} = (u, v)^\intercal$ of the light source pixel corresponding to the queried ray of sight. The renderer successively updates a data structure storing the light transport matrix $\mathbf{T}$ by increasing the entry corresponding to the obtained light source pixel $\mathbf{u}$ and the pixel sample $\mathbf{p}^{\mathrm{b}}$. Since for the inspection of transparent objects the entries of $\mathbf{T}$ are equal to zero for most camera pixels, data structures supporting sparsity (e.g., a sparse two-dimensional array) represent a suitable choice. Such arrays only require computer memory for elements with values different from zero.

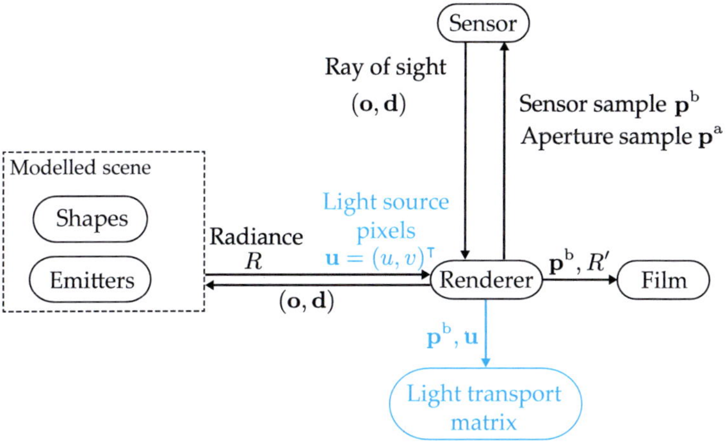

Figure 5.8: The modifications applied to the Mitsuba framework (drawn in blue) to support the simulation of light transport matrices.

5.3 Summary

This section introduced the Mitsuba renderer which allows to simulate physically realistic camera images of virtual scenes. After describing the main components of the rendering framework and their interaction, several plugins have been presented which have been added to the framework. Novel sensor plugins have been created: a plugin modeling the $4f$-light field camera introduced in Section 4.2.2 and a light field sensor plugin as it is required by the inverse light field illumination method explained in Chapter 7. A parallel emitter plugin has been presented for the simulation of collimated light sources. This plugin has also been extended to simulate spatially programmable light sources. As a counterpart of the light field sensor, a light field emitter plugin has been introduced allowing to emit four-dimensional light fields. Furthermore, the Mitsuba framework has been adequately extended to allow a streamlined simulation of light transport matrices. All of the simulated experiments reported in this thesis have been performed based on the Mitsuba framework and the introduced components (see Sections 6.1.3.1, 6.4.2 and 7.2.1).

6 Light field processing

This chapter presents methods for processing the light fields acquired by means of the acquisition approaches from Section 4. Section 6.1 covers the processing of light deflection maps by defining a spatial gradient for deflection maps and by introducing suitable distance functions. Section 6.2 introduces methods based on vector analysis for processing the signals of a position sensitive detector to reveal larger scattering defects. In Section 6.3, two image fusion approaches are described, which allow to learn and incorporate information about the test object to obtain inspection images where all of the nominal object features are suppressed and only the material defects remain visible. Section 6.4 shows how suitable features can be extracted from a light transport matrix that is acquired from a transparent test object to make scattering material defects visible. Section 6.5 closes the chapter with a summary.

6.1 Deflection map gradient

As introduced in Section 4.1, deflection maps represent light fields that are acquired under certain conditions. Deflection maps can either be acquired by illuminating the measurement field with collimated light rays and capturing a four dimensional light field (e.g., with the $4f$-light field camera or the laser deflection scanner) or by illuminating the measurement field with multiple collimated light beams that are tilted by different angles and capturing only those light rays propagating parallel to the optical axis (e.g., with the schlieren deflectometer). The idea of how to process deflection maps to reveal material defects inside transparent test objects has already been described in Section 1.2. Spatial discontinuities between adjacent measurement points with respect to the angular distribution of the captured light's direction of propagation can be used as an indicator of material defects or object edges.

In classical image processing of gray value images $g(\mathbf{x}), \mathbf{x} = (x, y)^\mathsf{T}$, jump discontinuities of the gray values of adjacent pixels represent so-called *edges*—that is, abrupt transitions between high and low gray values (or vice versa)— which are of high interest for many applications. Edges are manifested in the image signal by an extremum in the first derivative. Hence, these edges are commonly detected by means of the gradient of the gray values:

$$\text{grad}\; g(\mathbf{x}) = \begin{pmatrix} \frac{\partial g(\mathbf{x})}{\partial x} \\ \frac{\partial g(\mathbf{x})}{\partial y} \end{pmatrix} . \tag{6.1}$$

If there is an edge at location $\mathbf{x}$, then $\|\text{grad}\; g(\mathbf{x})\|$ will have a high value. An approximation of the gradient can be obtained by means of the so-called *symmetric difference quotient*

$$\text{grad}\; g(\mathbf{x}) \approx \begin{pmatrix} \frac{g(x+1,y)-g(x-1,y)}{2} \\ \frac{g(x,y+1)-g(x,y-1)}{2} \end{pmatrix} . \tag{6.2}$$

This idea can be adapted for the case of finding discontinuities in deflection maps with respect to the spatial coordinates. However, the gradient approximation formulated in Eq. (6.2) cannot be directly applied because the angular components of the deflection map represent two-dimensional gray value distributions instead of scalar gray values, as in the case of a common gray value image $g(\mathbf{x})$. Hence, Eq. (6.2) is reformulated as

$$\operatorname{grad} f(\mathbf{x}) \approx \begin{pmatrix} \frac{d\{f(x+1,y), f(x-1,y)\}}{2} \\ \frac{d\{f(x,y+1), f(x,y-1)\}}{2} \end{pmatrix}, \tag{6.3}$$

with $d\{\cdot,\cdot\}$ representing a suitable distance function and $f(\mathbf{x})$ an arbitrary function for which the gradient should be approximated. By setting $f(\mathbf{x}) := g(\mathbf{x})$ and $d\{k,l\} := k - l$, the two gradient approximations Eq. (6.2) and Eq. (6.3) are identical. For the application to a deflection map $a(\mathbf{m},\mathbf{j})$, Eq. (6.3) turns to

$$\operatorname{grad}_{\mathbf{m}} a(\mathbf{m},\cdot) \approx \frac{1}{2} \begin{pmatrix} d\{a\left((m+1,n)^\mathsf{T},\cdot\right), a\left((m-1,n)^\mathsf{T},\cdot\right)\} \\ d\{a\left((m,n+1)^\mathsf{T},\cdot\right), a\left((m,n-1)^\mathsf{T},\cdot\right)\} \end{pmatrix}. \tag{6.4}$$

Here, the distance function $d\{\cdot,\cdot\}$ has to process two two-dimensional intensity distributions, i.e.,

$$d\{a(\mathbf{m}_1,\cdot), a(\mathbf{m}_2,\cdot)\} : (\Omega_\mathrm{a} \to \mathcal{Q})^2 \to \mathbb{R}, \tag{6.5}$$

with Ω_a denoting the angular domain and $\mathcal{Q}$ the quantized intensity values of the deflection map $a(\mathbf{m},\mathbf{j})$. Since the angular components of the deflection maps can be interpreted as distributions of deflection angles at the respective spatial position, distance functions for comparing two-dimensional histograms represent a promising choice for the distance function $d\{\cdot,\cdot\}$. Further image processing and classification algorithms can then be applied to the norm of the gradient $\|\operatorname{grad}_{\mathbf{m}} a(\mathbf{m},\cdot)\|$, which represents a scalar value and thus can be handled as a gray value image.

As pointed out in the motivation of using light field methods given in the introduction (see Section 1.2), a suitable distance function should meet the following criteria:

1. **Sensitivity to strong shifts of peaks**: It should yield a high value, if a concentrated peak in the angular component of the first argument is strongly shifted compared to the angular component of the second argument, with the distance value being proportional to the peak distance.

2. **Sensitivity to spreadings of peaks**: It should yield a high value, if there is a concentrated peak in the angular component of the first argument and a broad intensity distribution lacking a distinct peak in the second argument (or vice versa).

3. **Sensitivity to intensity differences**: It should yield a value that is proportional to the difference of the overall intensities of the two arguments.

4. **Robustness against small variations**: It should yield a low value if the angular components of the two arguments show peaks at close deflection angles or similar intensity distributions in general.

5. **Low computational complexity**: Its computational complexity should be low.

The first two of these evaluation criteria are important for the detection of scattering defects which are mainly manifested in the angular distribution of the light's direction of propagation. The third criterion states that the distance function should be sensitive to intensity differences, so that light absorbing material defects can be made visible. The fourth criterion certifies that the surface curvature of the test object, which also leads to small differences of the deflection angles of adjacent light bundles, does not result in high distance values that could be ambiguous with respect to defects. The fifth statement requires the distance function to be efficiently computable, so that the inspection rates necessary for an industrial production line can be achieved.

Promising choices of distance functions that have been evaluated in this thesis are the so-called *Earth Mover's Distance* and the so-called *Generalized Cramér-von Mises Distance*, which are described in the following two sections 6.1.1 and 6.1.2. In Section 6.1.3, the two distances are evaluated in terms of several experiments.

6.1.1 Earth Mover's Distance

The Earth Mover's Distance (EMD) has its roots in transportation theory, where a similar concept was introduced in 1781 [Mon81] by Gaspard Monge. The EMD was first applied in the field of image processing in 1989 by Peleg et al. [Pel89], who used it to compare images of different resolutions.

The EMD represents a distance measure between probability density functions or normalized histograms that is based on the minimum costs for transforming a first histogram h_1 into a second histogram h_2 (or vice versa) by rearranging the probability mass. Formally, the EMD between two histograms h_1 and h_2 with N bins each can be expressed as the optimization problem:

$$\mathrm{EMD}(h_1, h_2) := \min_{\gamma(k,l)\in\mathcal{M}} \sum_{k=1}^{N} \sum_{l=1}^{N} \gamma(k,l)\, c(k,l)\,. \tag{6.6}$$

Here, $c(k,l)$ denotes the costs of moving one unit probability mass from bin k of h_1 to bin l of h_2 and the set

$$\begin{aligned}\mathcal{M} = \{\gamma(k,l) : &\gamma(k,l) \geq 0, \\ &\sum_{l=1}^{N} \gamma(k,l) = h_1(k), 1 \leq k \leq N, \\ &\sum_{k=1}^{N} \gamma(k,l) = h_2(l), 1 \leq l \leq N\}\end{aligned} \tag{6.7}$$

denotes the valid shifts of amounts of probability mass $\gamma(k,l)$ from bin k of h_1 to bin l of h_2. For one-dimensional histograms, the cost function $c(k,l)$ may canonically be chosen as the number of bins between bin k and bin l. However, for two- or higher-dimensional histograms, there are multiple choices for $c(\cdot,\cdot)$, such as the Euclidean distance (L_2-distance) or the Manhattan distance (L_1-distance). Depending on the distance function chosen for $c(\cdot,\cdot)$, the resulting computational complexity of the EMD can vary notably (see Section 6.1.1.1). Figure 6.1 shows an example calculation of the EMD for two one-dimensional histograms. Since the EMD is defined for normalized histograms, the deflection distributions contained in the deflection maps $a(\mathbf{m},\mathbf{j})$ have to be normalized to

$$\widetilde{a}(\mathbf{m},\mathbf{j}) = \frac{a(\mathbf{m},\mathbf{j})}{\sum\limits_{\mathbf{j}\in\Omega_\mathrm{a}} a(\mathbf{m},\mathbf{j})}\,, \tag{6.8}$$

before a gradient based on the EMD can be calculated.

6.1.1.1 Complexity analysis of the Earth Mover's Distance

The original formulation of the EMD using the Euclidean distance for $c(\cdot,\cdot)$ involves solving the optimization problem stated in Eq. (6.6), which can be done via linear programming. For deflection map processing, the EMD is applied to two-dimensional distributions of deflection angles. When assuming a square angular domain of size $|\Omega_{\mathrm{a}}| = n$, the resulting worst-case complexity is in $\mathcal{O}(n^3 \log n)$ for computing the corresponding linear program as shown in [Rub00]. For good-natured problem instances, the required computation time can be lower.

When choosing the L_1-norm for $c(\cdot, \cdot)$, the more efficient *Tree-EMD* algorithm introduced in [Lin07] can be employed. In this case a linear program still has to be solved but since the authors achieved to extensively reduce the number of unknown variables, they could empirically show that the average worst-case computational complexity is in $\mathcal{O}(n^2)$.

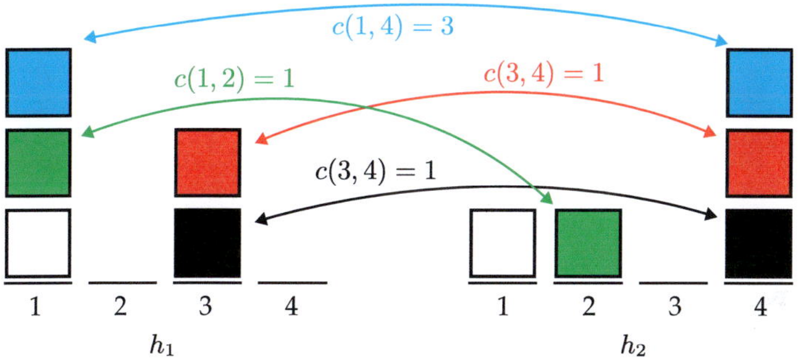

Figure 6.1: Illustration of calculating the EMD for two one-dimensional histograms h_1 and h_2 with 4 bins. To transform h_1 into h_2 with minimum costs, the blue unit mass has to be moved by three bins, the green, red and black unit masses have to be moved by one bin each and the white unit mass can remain at its bin. Hence, this results in $\mathrm{EMD}(h_1, h_2) = 3 + 1 + 1 + 1 = 6$.

6.1.1.2 Discussion of the Earth Mover's Distance

The following list discusses the EMD with respect to the criteria stated on p. 106 for a distance function $d\{\cdot, \cdot\}$:

1. **Sensitivity to strong shifts of peaks** – Criterion met: for two deflection distributions with distinct peaks, the corresponding probability masses would have to be moved over several bins resulting in a high EMD. Furthermore, since the EMD is proportional to the distance between the bins, the distance between the peaks would also be taken into account.
2. **Sensitivity to spreadings of peaks** – Criterion met: for a peak-shaped deflection distribution adjacent to a broad deflection distribution, the EMD would yield a high value since the probability mass concentrated in the peak would have to be moved to many different bins with partially large distances.
3. **Sensitivity to intensity differences** – Criterion not met: as the deflection distributions are normalized before being processed by the EMD, differences in the total intensity between the two deflection distributions are not taken into account and will not result in an increased EMD.
4. **Robustness against small variations** – Criterion met: since the EMD is sensitive to the distance by which the intensities, i.e., the probability masses, have to be moved, two close peaks or two similar deflection distributions result in a low EMD.
5. **Low computational complexity** – Criterion partially met: the high computational complexity of $\mathcal{O}(n^3 \log n)$ of the conventional, L_2-based EMD will result in processing times that are intolerable for most industrial production lines. However, for light field, particularly deflection map acquisition systems with a low angular resolution, the computational complexity of $\mathcal{O}(n^2)$ of the Tree-EMD algorithm can be sufficient.

Although it does not meet all of the listed criteria, the EMD still represents a suitable distance function due to its interpretability.

6.1.2 Generalized Cramér-von Mises Distance

In different fields, e.g., probability theory or content based image retrieval systems, applications and algorithms require to compute a distance measure between two random vectors, respectively, between their corresponding probability density functions to assess their similarity [Cha07]. If one of the two random vectors is of discrete type, then it is a possible way to use the cumulative distributions of the probability density functions for further processing steps. However, cumulative distributions are only well defined for one-dimensional random vectors and not for the multivariate case. In [Han08], Hanebeck et al. introduce the so-called Generalized Cramér-von Mises Distance (CMD) by extending the existing Cramér-von Mises Distance by so-called *localized cumulative distributions (LCDs)*. As will be shown later in this section, the LCD of a random vector is obtained via a rectangular kernel transform of the underlying probability density function. A generalized formulation of the Cramér-von Mises Distance is derived by means of the LCDs of two random vectors, which can be of continuous or discrete type. As detailed later, unfortunately the computation of the LCD and the CMD of two discrete two-dimensional random vectors $\mathbf{x}, \mathbf{y} \in \Omega_{\mathrm{a}}$, as would be the case for processing deflection maps with an angular domain of $|\Omega_{\mathrm{a}}| = n$, would have a computational complexity of $\mathcal{O}(n^{2.5})$, which might result in processing times that are too high for industrial inspection applications. However, in Section 6.1.2.2 this thesis introduces *fastCMD*, which is a fast algorithm for computing the CMD that has a reduced computational complexity of $\mathcal{O}(n^{1.5})$.

Localized cumulative distributions As mentioned previously, the conventional formulation of cumulative distributions is not well-defined for the multivariate case. The localized cumulative distributions (LCDs) introduced in [Han08] resolve this issue as follows. Let $\mathbf{x} \in \mathbb{R}^N, N \in \mathbb{N}$ be a random vector and let $f : \mathbb{R}^N \to \mathbb{R}_+$ denote the corresponding probability density function. The respective LCD $F(\mathbf{x}, \mathbf{b})$ is defined as

$$F(\mathbf{x}, \mathbf{b}) := P(|\mathbf{x} - \mathbf{x}| \leq \mathbf{b}) \,, \tag{6.9}$$

$$F(\cdot, \cdot) : \Omega \to [0, 1] \,,$$

$$\Omega \subset \mathbb{R}^N \times \mathbb{R}_+^N, \ \mathbf{b} \in \mathbb{R}_+^N \,, \tag{6.10}$$

with $\mathbf{x} \leq \mathbf{y}$, $\mathbf{x}, \mathbf{y} \in \mathbb{R}_+^N$ representing a component-wise relation that holds only if $\forall j \in [1, \ldots, N] : x_j \leq y_j$. Via the probability density function $f(\mathbf{x})$ corresponding to $\mathbf{x}$, the LCD $F(\mathbf{x}, \mathbf{b})$ can be calculated as

$$F(\mathbf{x}, \mathbf{b}) = \begin{cases} \int\limits_{\mathbf{x}-\mathbf{b}}^{\mathbf{x}+\mathbf{b}} f(\mathbf{t}) \mathrm{d}\mathbf{t} & \text{, if } \mathbf{x} \text{ continuous,} \\ \sum\limits_{\mathbf{t}=\max\{\mathbf{0}, \mathbf{x}-\mathbf{b}\}}^{\min\{\mathbf{x}_{\max}, \mathbf{x}+\mathbf{b}\}} f(\mathbf{t}) & \text{, if } \mathbf{x} \text{ discrete,} \end{cases} \tag{6.11}$$

with $\mathbf{x}_{\max}$ denoting the upper limit of the domain of $\mathbf{x}$, $\max\{\mathbf{x}\}$, $\min\{\mathbf{x}\}$ denoting element-wise maximum and minimum operators and $\mathbf{0} = (0, \ldots, 0)^\intercal$ denoting the zero vector. Due to the integration, respectively, the summation of the probability density function inside the boundaries specified by $\mathbf{b}$, the LCD $F(\mathbf{x}, \mathbf{b})$ can be imagined as an integral transform with a rectangular kernel $\mathbf{b}$. In contrast to the conventional formulation of cumulative distributions, the LCD of a multivariate random vector is unique and well-defined. As for the application of processing deflection maps especially discrete variables are concerned, the discrete case of Eq. (6.11) is implicitly considered from now on.

Generalized Cramér-von Mises Distance Based on the definition of the LCD, the Generalized Cramér-von Mises Distance can now be introduced. Let $\mathbf{x}, \mathbf{y} \in \mathbb{R}^N, N \in \mathbb{N}$ be two random vectors, let $f(\mathbf{x}), h(\mathbf{x})$ be their corresponding probability density functions and let $F(\mathbf{x}, \mathbf{b}), H(\mathbf{x}, \mathbf{b})$ denote their LCDs (as defined previously). For the continuous case, the CMD of $f(\mathbf{x})$ and $h(\mathbf{x})$ is given by:

$$\mathrm{CMD}(f, h) := \int_{\mathbb{R}^N} \int_{\mathbb{R}_+^N} \left(F(\mathbf{x}, \mathbf{b}) - H(\mathbf{x}, \mathbf{b})\right)^2 \mathrm{d}\mathbf{b}\mathrm{d}\mathbf{x} \,. \tag{6.12}$$

For discrete random vectors, the integrals turn into summations yielding:

$$\mathrm{CMD}(f, h) = \sum_{\mathbf{x} \in \Omega} \sum_{b=0}^{b_{\max}} \left(F(\mathbf{x}, (b, \ldots, b)^{\intercal}) - H(\mathbf{x}, (b, \ldots, b)^{\intercal})\right)^2 \,, \tag{6.13}$$

with Ω denoting the domain of the probability density functions and $b_{\max}$ representing the absolute maximum component value of Ω, i.e., the maximum kernel size necessary to capture the whole probability density functions.

6.1.2.1 Complexity analysis of the naive calculation of the Generalized Cramér-von Mises Distance

In the following, the computational complexity of the naive calculation of the CMD is analyzed. As in the context of this thesis, the CMD is employed to compare deflection distributions of deflection maps, the complexity analysis is restricted to two-dimensional deflection distributions. The angular domain is assumed to be $|\Omega_{\mathrm{a}}| = n$ with equal resolutions in horizontal and vertical direction, resulting in a maximum kernel size of $b_{\max} = \sqrt{n}$.

As a first step, the complexity of one evaluation of the LCD is derived. Since quadratic kernels are used in the definition of the CMD in Eq. (6.13) (i.e., the kernel vector $\mathbf{b} = (b,b)^\intercal$ contains the same value b for both dimensions), evaluating the LCD $F(\mathbf{x},\mathbf{b})$ involves b^2 summations. Hence, the complexity of one evaluation of the LCD is $\mathcal{O}(b^2)$.

Each iteration of the inner summation of the CMD in Eq. (6.13) also has a complexity of $\mathcal{O}(b^2)$ because it involves two LCD evaluations. Since b is increased from $b = 0$ to $b = b_{\max} = \sqrt{n}$, the inner summation of Eq. (6.13) involves a number of

$$\mathcal{O}\left(\sum_{b=0}^{\sqrt{n}} b^2\right) = \mathcal{O}\left(\frac{n^{1.5}}{3} + \frac{n}{2} + \frac{\sqrt{n}}{6}\right) = \mathcal{O}(n^{1.5}) \tag{6.14}$$

operations according to Faulhaber's formula [Abr13] and, hence, has a complexity of $\mathcal{O}(n^{1.5})$. Since these operations are performed n times by the outer summation of Eq. (6.13), the naive calculation of the CMD has a total complexity of $\mathcal{O}(n^{2.5})$:

$$\mathrm{CMD}(f,h) = \underbrace{\sum_{\mathbf{x}\in\Omega} \underbrace{\sum_{b=0}^{b_{\max}} \overbrace{\left(F(\mathbf{x},(b,\ldots,b)^\intercal) - H(\mathbf{x},(b,\ldots,b)^\intercal)\right)^2}^{\in\mathcal{O}(b^2)}}_{\in\mathcal{O}(n^{1.5})}}_{\in\mathcal{O}(n^{2.5})} . \tag{6.15}$$

6.1.2.2 Fast Generalized Cramér-von Mises Distance

This section introduces the fastCMD-algorithm [Mey17a] which accelerates the computation of the discrete CMD by adequately employing the concept of summed area tables (SATs). After explaining

SATs, this section covers the fastCMD-algorithm and derives its computational complexity.

Summed area tables A SAT is a data structure that is precomputed for a two-dimensional input array and allows to obtain the sum of the array entries inside any arbitrary rectangular region in constant time [Cro84]. The two-dimensional input array is given by $i(m,n) \in \mathbb{R}, m \in [1,\ldots,M], n \in [1,\ldots,N]$. The corresponding SAT $\mathfrak{s}$ is then defined as:

$$\mathfrak{s}(m,n) := \begin{cases} 0 & \text{if } \min\{m,n\} \leq 1\,, \\ \sum\limits_{m_\mathrm{f}=1}^{m} \sum\limits_{n_\mathrm{f}=1}^{n} i(m_\mathrm{f}, n_\mathrm{f}) & \text{otherwise.} \end{cases} \tag{6.16}$$

The calculation of the SAT $\mathfrak{s}$ can be performed in a single sweep over the input array i by means of the following iterative formulation:

$$\mathfrak{s}(m,n) = i(m,n) + \mathfrak{s}(m-1,n) + \mathfrak{s}(m,n-1) - \mathfrak{s}(m-1,n-1)\,. \tag{6.17}$$

With the help of $\mathfrak{s}$, the sum of the array entries of i inside a rectangular region given by $m \in [m_\mathrm{f}, m_\mathrm{t}]$, $n \in [n_\mathrm{f}, n_\mathrm{t}]$ can be obtained with only three arithmetic operations performed on $\mathfrak{s}$:

$$\begin{aligned} \sum_{m=m_\mathrm{f}}^{m_\mathrm{t}} \sum_{n=n_\mathrm{f}}^{n_\mathrm{t}} i(m,n) &= \mathfrak{s}(m_\mathrm{t}, n_\mathrm{t}) - \mathfrak{s}(m_\mathrm{f}-1, n_\mathrm{t}) \\ &\quad - \mathfrak{s}(m_\mathrm{t}, n_\mathrm{f}-1) + \mathfrak{s}(m_\mathrm{f}-1, n_\mathrm{f}-1)\,. \end{aligned} \tag{6.18}$$

The iterative formulation stated in Eq. (6.17) for creating the SAT involves only seven constant time arithmetic operations and four constant time array accesses which have to be performed for each array element of i, that is, a total of $M \cdot N$ times. Hence, constructing $\mathfrak{s}$ has a computational complexity of $\mathcal{O}(M \cdot N)$. By means of Eq. (6.18), obtaining the sum of entries of i in any arbitrary rectangular region has a complexity of $\mathcal{O}(1)$. Figure 6.2 shows an illustration of the scheme of Eq. (6.18).

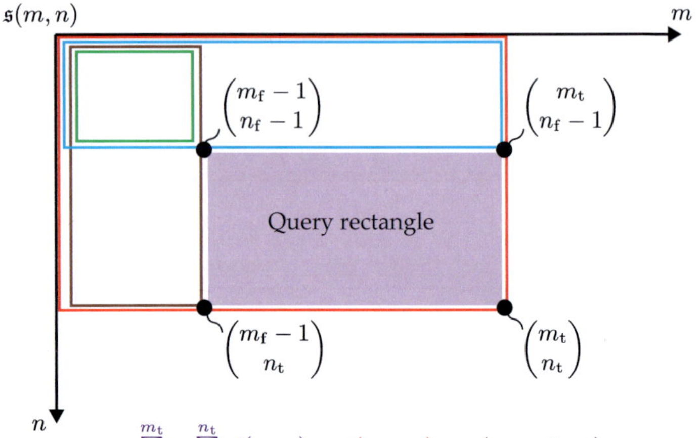

$$\sum_{m=m_\mathrm{f}}^{m_\mathrm{t}} \sum_{n=n_\mathrm{f}}^{n_\mathrm{t}} i(m,n) = \mathfrak{s}(m_\mathrm{t}, n_\mathrm{t}) - \mathfrak{s}(m_\mathrm{f}-1, n_\mathrm{t}) - \mathfrak{s}(m_\mathrm{t}, n_\mathrm{f}-1) + \mathfrak{s}(m_\mathrm{f}-1, n_\mathrm{f}-1)$$

Figure 6.2: Using a summed area table for calculating the sum of the array entries inside the query rectangle $m \in [m_\mathrm{f}, m_\mathrm{t}]$, $n \in [n_\mathrm{f}, n_\mathrm{t}]$ (purple area): from the red component $\mathfrak{s}(m_\mathrm{t}, n_\mathrm{t})$, the brown component $\mathfrak{s}(m_\mathrm{f}-1, n_\mathrm{t})$ and the blue component $\mathfrak{s}(m_\mathrm{t}, n_\mathrm{f}-1)$ are subtracted. The sum of the green region has now been subtracted twice, which is compensated by adding $\mathfrak{s}(m_\mathrm{f}-1, n_\mathrm{f}-1)$.

FastCMD The main idea of fastCMD lies in accelerating Eq. (6.11), that is, the calculation of the LCDs by means of summed area tables. Since the discrete case of Eq. (6.11) represents a summation over a rectangular region of a two-dimensional data structure, a summed area table can be precomputed for $f(\mathbf{m})$ to speed up the evaluation of $F(\mathbf{m}, \mathbf{b})$. The pseudocode for the fastCMD-algorithm is listed in

Algorithm 6.1: The fastCMD-algorithm for calculating the CMD of two discrete probability density functions $f(\mathbf{m}), h(\mathbf{m})$.

fastCMD$\big(f(\mathbf{m}), h(\mathbf{m})\big)$
$\mathfrak{f} \leftarrow$ generateSummedAreaTable$(f(\mathbf{m}))$
$\mathfrak{h} \leftarrow$ generateSummedAreaTable$(h(\mathbf{m}))$
CMD $\leftarrow 0$
$b_{\max} = \max\{M, N\}$
for $m = 1, \ldots, M$ **do**
  for $n = 1, \ldots, N$ **do**
    for $b = 0, \ldots, b_{\max}$ **do**
      $\text{CMD} \leftarrow \text{CMD} + \big(\text{fastLCD}(\mathfrak{f}, (m,n)^\mathsf{T}, (b,b)^\mathsf{T}) - \text{fastLCD}(\mathfrak{h}, (m,n)^\mathsf{T}, (b,b)^\mathsf{T})\big)^2$
    end for
  end for
end for
return CMD

Algorithm 6.1. The algorithm requires two discrete probability functions $f(\mathbf{m}), h(\mathbf{m})$ as input parameters. First, it calculates the SAT data structures $\mathfrak{f}, \mathfrak{h}$ for $f(\mathbf{m})$ and $h(\mathbf{m})$ according to Eq. (6.17). The following two outer for-loops iterate over the domain $\Omega = [1, \ldots, M] \times [1, \ldots, N]$ and the third, inner for-loop iterates over the kernel sizes of $b \in [0, \ldots, b_{\max}]$, with $b_{\max} = \max\{M, N\}$, and updates the calculation of the CMD. Here, the fastLCD-algorithm listed in Algorithm 6.2 is called and passed the precomputed SATs $\mathfrak{f}, \mathfrak{h}$, the requested position $(m, n)^\mathsf{T}$ and kernel sizes $(b, b)^\mathsf{T}$. The fastLCD-algorithm evaluates the LCD according to Eq. (6.18).

6.1.2.3 Complexity analysis of the fastCMD-algorithm

For the complexity analysis of the fastCMD-algorithm, the application of deflection map processing is again considered. Therefore, the discrete two-dimensional deflection distributions are assumed as input probability density functions with square-shaped angular domains of $|\Omega_\mathrm{a}| = n$. The resulting maximum kernel size is $b_\mathrm{max} = \sqrt{n}$.

Precomputing the SATs has a complexity of $\mathcal{O}(n)$ because they can be generated by means of a single sweep over the probability density functions. The inner loop of the three nested for-loops iteratively updates the calculated CMD by computing the difference between two evaluations of the *fastLCD*-algorithm shown in Algorithm 6.2. Since one evaluation of fastLCD has a constant complexity of $\mathcal{O}(1)$, the inner loop of fastCMD also has a complexity of $\mathcal{O}(1)$. Since all three loops run for $\sqrt{n}$ iterations each, fastCMD has a total complexity of $\mathcal{O}(1 \cdot (\sqrt{n})^3) = \mathcal{O}(n^{1.5})$. An empirical comparison of the runtimes of the fastCMD-algorithm and the direct computation of the CMD is provided in Appendix A.2.

Algorithm 6.2: Algorithm for evaluating the localized cumulative distribution $F((m,n)^\intercal, (b,b)^\intercal)$ of a probability density function based on its summed area table $\mathfrak{s}$ at position $(m,n)^\intercal$ with kernel sizes $(b,b)^\intercal$.

fastLCD$(\mathfrak{s}, (m,n)^\intercal, (b,b)^\intercal)$
$m_\mathrm{f} \leftarrow \max\{0, m-b\}$
$m_\mathrm{t} \leftarrow \min\{M, m+b\}$
$n_\mathrm{f} \leftarrow \max\{0, n-b\}$
$n_\mathrm{t} \leftarrow \min\{N, n+b\}$
return $\mathfrak{s}(m_\mathrm{t}, n_\mathrm{t}) - \mathfrak{s}(m_\mathrm{f}-1, n_\mathrm{t}) - \mathfrak{s}(m_\mathrm{t}, n_\mathrm{f}-1) + \mathfrak{s}(m_\mathrm{f}-1, n_\mathrm{f}-1)$

6.1.2.4 Discussion of the Generalized Cramér-von Mises Distance

The following list discusses the suitability of the CMD as a distance function $d\{\cdot,\cdot\}$ with respect to the criteria stated on p. 106.

1. **Sensitivity to strong shifts of peaks** – Criterion met: for two deflection distributions with distinct peaks, the two LCDs are only equal to each other for larger kernel sizes, i.e., b-values.

Hence, all smaller kernels will result in non-zero differences between the LCD-evaluations of the two deflection distributions in Eq. (6.13) and, consequently, yield a high CMD. Since the distance between the two peaks determines the kernel size from which the two LCDs will be equal, the CMD is also proportional to the peak distance.

2. **Sensitivity to spreadings of peaks** – Criterion met: if a deflection distribution with an intensity peak is adjacent to a deflection distribution with a broad intensity profile, then the resulting CMD will be high. This happens because the LCDs for the peak-shaped distribution will have a jump discontinuity from low to high as soon as the integration kernel overlaps with the peak and the LCDs of the second deflection distribution will increase gradually for increasing kernel sizes as the intensity is spread over most of the whole angular domain.

3. **Sensitivity to intensity differences** – Criterion met: although the CMD has originally been formulated for probability density functions, it does not require its input arguments to be normalized. Two deflection distributions that are identical except for a scaling factor (i.e., which have different total intensities) will result in different LCD-values for each kernel size and will, therefore, have a non-zero CMD.

4. **Robustness against small variations** – Criterion met: for two deflection distributions with close peaks, the LCDs will only have different values for few locations and few kernel sizes and, therefore, the resulting CMD will be low.

5. **Low computational complexity** – Criterion met: as shown in Section 6.1.2.2, the CMD can be calculated by means of the

fastCMD-algorithm, which has a complexity of $\mathcal{O}(n^{1.5})$ for an angular domain of $|\Omega_\mathrm{a}| = n$.

The Generalized Cramér-von Mises Distance meets all the postulated criteria and, hence, represents a promising choice for a distance function $d\{\cdot,\cdot\}$.

6.1.3 Experiments

This section covers the conducted experiments with respect to the processing of deflection maps. In the performed experiments, most of the acquisition systems described in Chapter 4 have been used to capture deflection maps of transparent test objects. Section 6.1.3.1 describes the experiments based on simulated light field images and Section 6.1.3.2 covers the experiments involving real sensor data. For the acquired deflection maps, the gradient of Eq. (6.4) has been calculated using both the Earth Mover's Distance (see Section 6.1.1) and the Generalized Cramér-von Mises Distance (see Section 6.1.2). The resulting images show the gradient's norm $\|\mathrm{grad}_\mathbf{m}\ a(\mathbf{m},\cdot)\|$ in pseudo colors. All of the experiments include a comparison to a conventional inspection image of the respective test object. Since all considered test objects represent uncooperative test objects, a pure bright field setup has been employed for the comparison because a dark field setup would have to be individually optimized. For a quantitative evaluation, the *peak contrast-to-noise ratio* CNR is determined based on the resulting inspection images [Des10]. The CNR is calculated for every considered defect of interest:

$$\mathrm{CNR} = \frac{|\tilde{s} - \hat{\mu}|}{\hat{\sigma}}\,, \tag{6.19}$$

with $\tilde{s}$ denoting the maximum image value of the image region covered by the defect, $\hat{\mu}$ denoting the mean value and $\hat{\sigma}$ the standard deviation estimated for a defect-free image region near the defect. The CNR measures to what extent a defect emerges out of the background noise. According to the so-called *Rose criterion* [Ros73], a CNR of at least 5 is required to enable a reliable detection of defects.

6.1.3.1 Simulated experiments

The experiments described in this section are based on simulated light field images of the $4f$-light field camera because no prototype currently exists of this sensor system. The light field images have been simulated using the rendering framework Mitsuba, in concert with the plugins introduced in Chapter 5.

$4f$-light field camera For the following experiments, a double-convex glass lens was used as the test object. As stated in the introduction (see Section 1.1.1), such lenses represent uncooperative test objects. In the simulated scene, the test object has been illuminated with a collimated light bundle from one side and has been imaged with the $4f$-light field camera from its other side, so that deflection maps could be acquired. The $4f$-light field camera had been parametrized with a spatial resolution of 555×555 and an angular resolution of 9×9 resulting in a sensor with $(555 \cdot 9)^2 = 24,950,025$ pixels. For a physical realization, these settings would require a 555×555 micro-lens array and a 25 mega pixel camera sensor, which represent feasible parameters. The test object has been artificially affected by two different material defects. On the one hand, parts of a small sphere have been removed from the test object's surface to represent an incomplete shape. On the other hand, a small sphere with the refraction index of air has been introduced into the test object's center to simulate an enclosed air bubble. The detection of the latter defect is especially challenging for conventional inspection systems. For comparison, images of such a system have been obtained by simulating a telecentric camera and an area illumination.

Figure 6.3 shows the resulting inspection images. The following table lists the CNRs calculated according to Eq. (6.19) for the two defects:

	CNR for	
Method	Shape defect	Scattering inclusion
EMD	6.64	4.87
CMD	6.94	**6.67**
Telecentric camera	**136.33**	0.22

(The highest CNR of every column is set in bold.)

In the images of all three inspection approaches, the first test object can be identified as being free from defects. However, the conventional inspection system yields the clearest image, whereas there is some intense noise present in the visualizations of the gradient norms. This noise is caused by rendering artifacts. The shape defect (i.e., the incomplete test object surface) is made visible by all candidates but with most contrast by the conventional imaging system that achieves a CNR of 136.33. The enclosed air bubble is only revealed by the light field approaches and not by the telecentric camera system. Due to the scattering nature of this defect, it is mainly manifested in the distribution of the light's direction of propagation. In contrast to the conventional inspection system, the $4f$-light field camera is able to capture this information and the EMD and the CMD clearly show the resulting discontinuities. With a CNR of 6.67, the CMD is superior to the EMD with a CNR of 4.87.

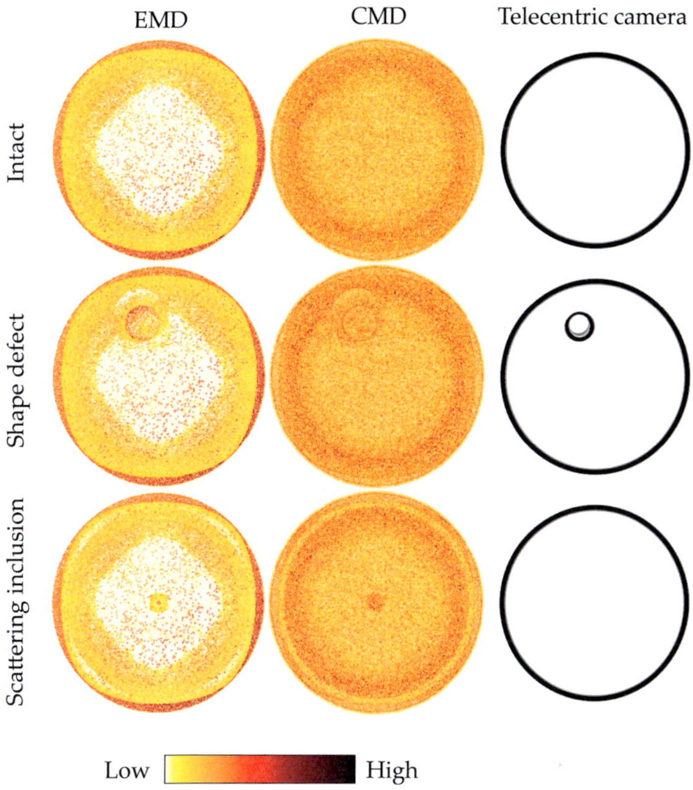

Figure 6.3: Inspection images of the simulated experiments. The first two columns show pseudo color visualizations of the norm of the gradient of the $4f$-light field camera's images by employing the EMD and the CMD, respectively. The third column shows the images of the telecentric camera system. The first row corresponds to a defect-free test object and the remaining two rows correspond to the test objects affected by material defects.

The processing[1] of the deflection maps took approximately 10.5 s $\pm$ 0.2 s for both the EMD and the CMD. One would expect that the time required for computing the EMD is longer than for the CMD.

[1] Technical details of the employed computer system are provided in Appendix A.3.

Since the Tree-EMD's complexity of $\mathcal{O}(n^2)$ represents the worst-case, faster computation times are possible.

6.1.3.2 Real experiments

This section describes the experiments performed with real imaging systems. Prototypes of the schlieren deflectometer and the laser deflection scanner have been used to acquire deflection maps of different kinds of test objects. These deflection maps are processed by computing the gradient from Eq. (6.4) based on the EMD and the CMD. The resulting gradients are visualized by showing their norm in pseudo colors. For comparison, all of the inspected test objects have also been imaged with conventional machine vision components.

Glass gob imaged by schlieren deflectometer As mentioned in the introduction (see Section 1.1.1), cylindrically shaped preforms made out of glass, so-called gobs, represent important intermediate products for the glass industry (see Fig. 1.2 and Fig. 1.3). Due to their geometry, gobs represent uncooperative test objects. A sample gob affected by enclosed absorbing particles and enclosed scattering bubbles has been used as a test object. A prototype of the schlieren deflectometer (see Section 4.3.2) with a spatial resolution of 1210×1210 and an angular resolution of 9×9 has been used to acquire deflection maps of the test object. To evaluate whether the usage of the EMD and the CMD, which represent rather complex distance functions, is at all necessary, a more simple norm based on the Euclidean distance has also been employed in the scope of these experiments.

The corresponding distance function is given by

$$d_{\mathrm{E}}\{a(\mathbf{m}_1,\cdot),a(\mathbf{m}_2,\cdot)\} = \sqrt{\sum_{\mathbf{j}\in\Omega_{\mathrm{a}}} \left(a(\mathbf{m}_1,\mathbf{j}) - a(\mathbf{m}_2,\mathbf{j})\right)^2}\,. \tag{6.20}$$

Figure 6.4 shows the resulting inspection images for the performed experiments. The image regions corresponding to the material defects are highlighted by colored rectangles. The following table lists the CNRs calculated according to Eq. (6.19) for the three numbered regions:

	CNR for region		
Method	1	2	3
EMD	1.88	7.65	3.81
CMD	**23.38**	**15.07**	**14.28**
Euclidean	3.84	5.09	5.41
Conventional bright field	1.83	1.05	0.46

(The highest CNR of every column is set in bold.)

The test object is affected by several absorbing contaminants (highlighted in red) and scattering air bubbles (highlighted in blue). The CMD clearly shows the contained material defects, whereas they only appear with low contrast or not at all in the image of the EMD. This is also supported by the calculated CNRs which are lower for the EMD compared to the CMD. Due to the normalization, the EMD achieves the lowest CNR for the defect in region 1 since it is not sensitive to absorbing defects which do not cause any scattering. The con-

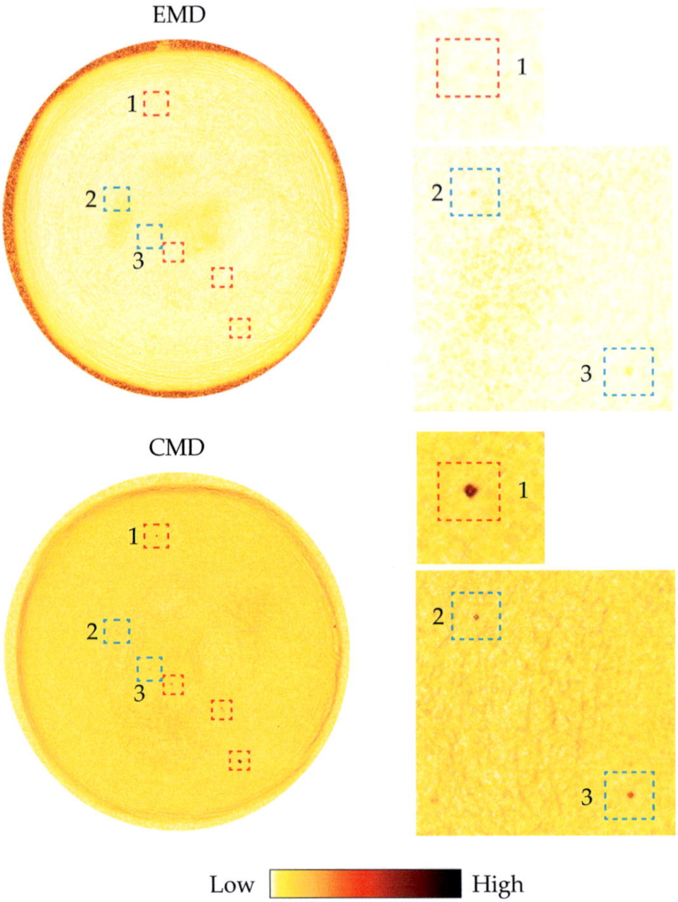

Figure 6.4: Inspection images for the investigated gob. The results of the gradient-based approaches are shown in pseudo colors. Some of the defects are magnified on the corresponding right-hand sides.

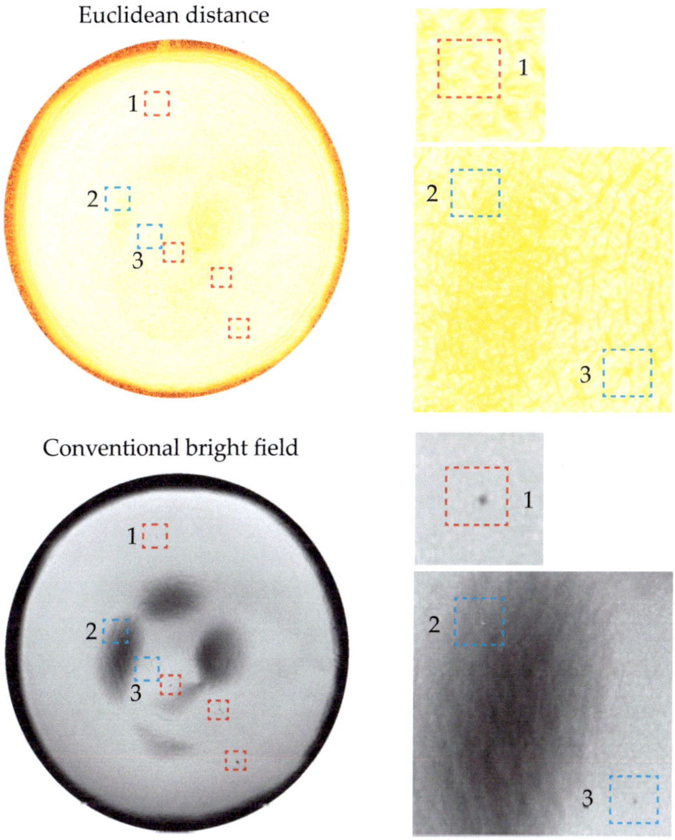

(Figure 6.4 continued)

ventional bright field setup reveals some of the absorbing defects but still with lower contrast compared to the CMD. The enclosed scattering air bubbles are revealed best by the CMD. Furthermore, the conventional bright field image is impaired by three large shadow-like artifacts that are caused by the gob's complex geometry. In the inspection image obtained by employing the Euclidean distance in the gradient formulation, the defects are only barely or not at all visible what justifies the usage of complex distance functions.

Acquiring the deflection maps required ca. 3 s and their processing took ca. $61\ \mathrm{s} \pm 1\ \mathrm{s}$ for the EMD and $50\ \mathrm{s} \pm 2\ \mathrm{s}$ for the CMD. These high computation times result from the high spatial resolution in combination with the angular resolution of 9×9.

Automotive headlamp cover imaged with a laser deflection scanner A prototype of the laser deflection scanner introduced in Section 4.4.1 has been employed to acquire deflection maps of an automotive headlamp cover made of transparent plastic. Figure 6.5 shows photographs of such headlamp covers. As can be seen, test objects of these types have a very complex geometry and also show many design features. Hence, these test objects clearly belong to the group of uncooperative test objects. Important types of material defects that have to be detected are light absorbing contaminants, scattering impurities (e.g., air bubbles) and scratches. The spatial resolution of the employed laser deflection scanner is defined by the length of its scan line of 1397 pixels. It supports an angular resolution of 3×3. Figure 6.6 shows the resulting inspection images. Besides the EMD-based and CMD-based deflection map gradients, a bright field acquisition of a conventional entocentric camera system is also shown. The image regions corresponding to the material defects are highlighted by colored rectangles.

(a)

(b)

Figure 6.5: Photographs of two examples of plastic headlamp covers: (a) mounted in front of the headlamps of an automobile (highlighted in blue); (b) the test object used in the experiments.

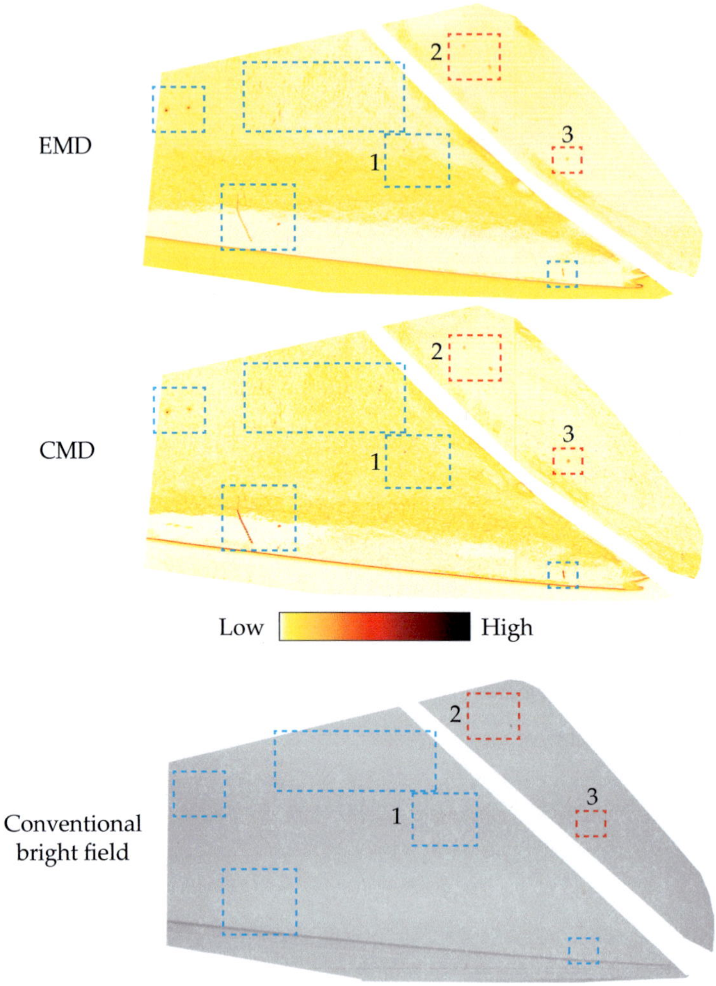

Figure 6.6: Inspection images for the investigated headlamp cover. Intense edges and design structures of the test object have been masked out. The results of the gradient-based approaches are shown in pseudo colors. Some of the defects are magnified on the right-hand side.

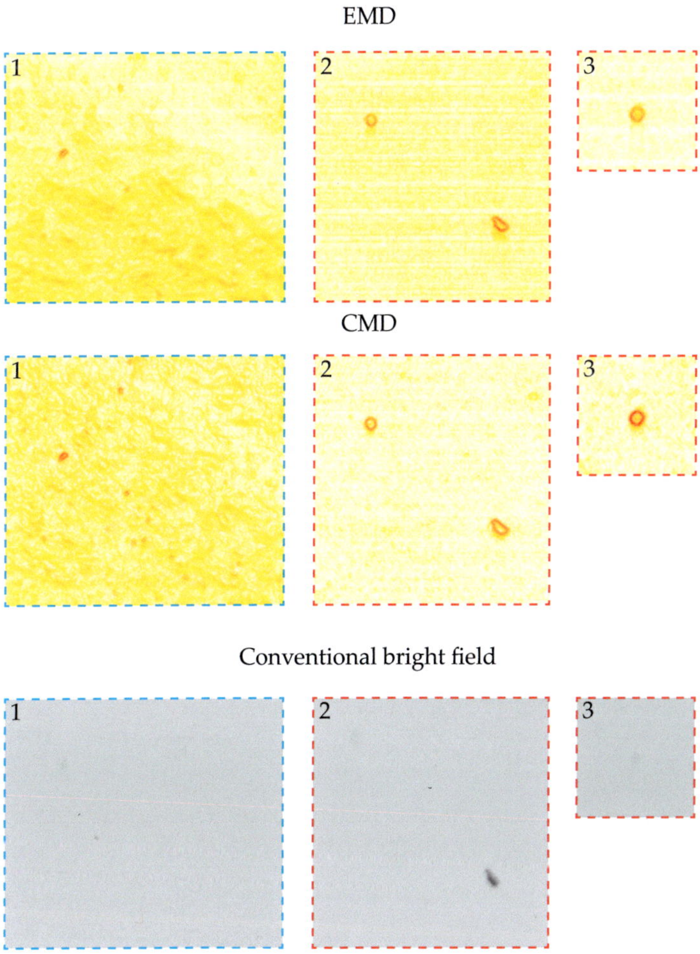

(Figure 6.6 continued)

The following table lists the CNRs calculated according to Eq. (6.19) for the three numbered regions:

	CNR for region		
Method	1	2	3
EMD	12.29	30.73	30.22
CMD	**23.04**	**31.11**	**40.91**
Conventional bright field	4.29	22.86	4.29

(The highest CNR of every column is set in bold.)

The absorbing defects (located in the red rectangles, regions 2 and 3) are made visible by all gradient-based approaches and the lower right-hand defect in region 2 can slightly be seen in the camera image. The upper left-hand defect in this region absorbs only little light but leads to some scattering which is why it is revealed by the gradient-based approaches but not by the bright field setup. These observations are also supported by the CNRs. The scattering defects, which are highlighted by the blue rectangles, are clearly visible in the results of the EMD and the CMD but are barely or not at all visible in the image of the conventional inspection system. The scattering defects inside region 1 are revealed with more contrast by the CMD with a CNR of 23.04 compared to the EMD with a CNR of 12.29 and one of them is faintly observable in the conventional bright field image.

The current prototype of the laser deflection scanner required ca. 5 s for capturing the deflection maps. The processing took approximately 10.5 s $\pm$ 0.5 s for the EMD and 2.6 s $\pm$ 0.2 s for the CMD.

Washing machine door glass imaged with laser deflection scanner For a further experiment, deflection maps of a washing machine door glass have been acquired with the same laser deflection scanner. As can be seen, such objects also represent uncooperative test objects due to their complex geometry. Especially important material defects for these test objects are sharp-edged surface defects that could damage the laundry inside the washing machine. In addition to shape defects, the inspected test object is affected by some light scattering impurities and by scratches.

Figure 6.7 shows the resulting inspection images and the following table lists the CNRs computed according to Eq. (6.19) for the highlighted regions 1, 2 and 3:

	CNR for region		
Method	1	2	3
EMD	**11.59**	26.90	34.86
CMD	11.55	**41.77**	**40.41**
Conventional bright field	3.76	3.09	3.09

(The highest CNR of every column is set in bold.)

The scratch inside region 1 is clearly imaged by both gradient-based approaches and leads to nearly the same CNRs for the EMD and the CMD. The scattering defects shown in region 2 are revealed by the EMD and the CMD, with the CMD achieving the highest CNR. This also applies for the important shape defect in region 3. The conventional bright field image does not reveal any of the material defects.

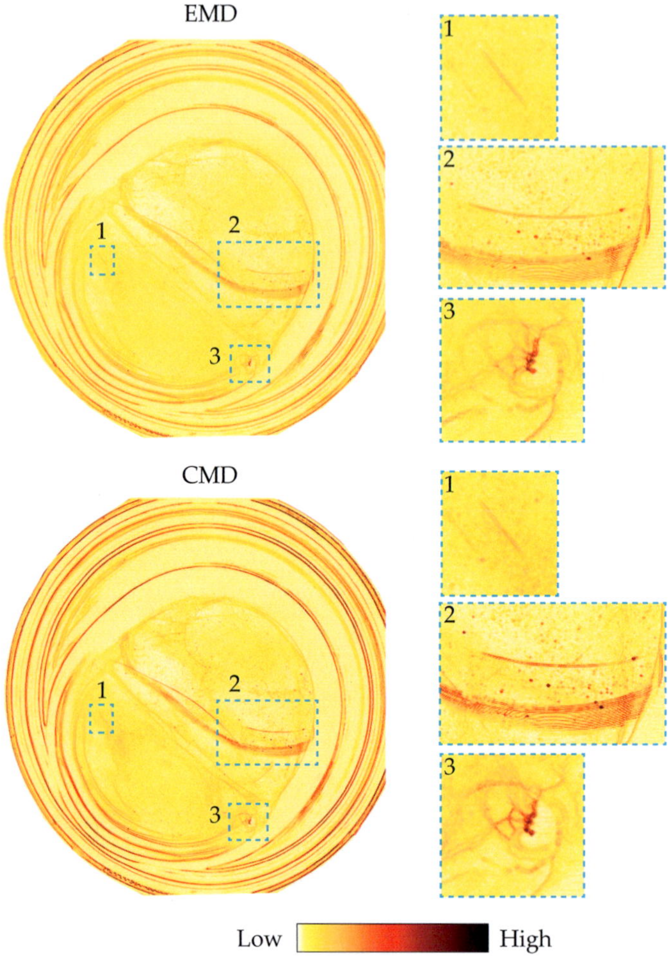

Figure 6.7: Inspection images for the investigated washing machine door glass. The results of the gradient-based approaches are shown in pseudo colors. Some of the defects are magnified on the corresponding right-hand sides.

Acquiring the deflection maps required ca. 5 s and their processing took approximately $7.4 \text{ s} \pm 0.3 \text{ s}$ for the EMD and $2.23 \text{ s} \pm 0.05 \text{ s}$ for the CMD.

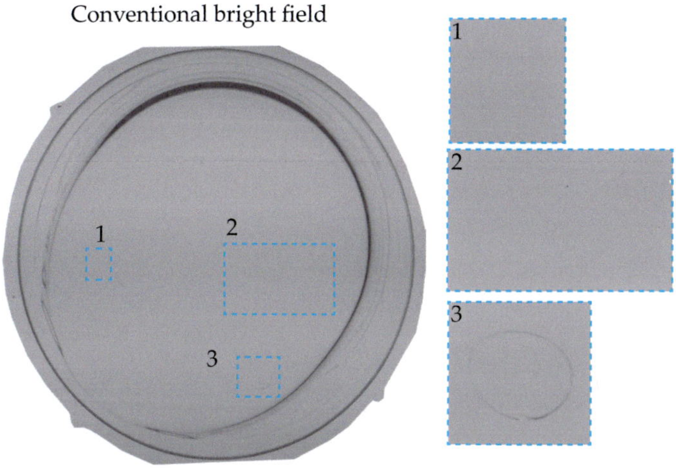

(Figure 6.7 continued)

6.1.4 Discussion

The experiments showed that acquiring and processing light deflection maps is a suitable approach for inspecting complex-shaped transparent objects. Whether the schlieren deflectometer or the laser deflection scanner should be chosen for the acquisition depends on the size of the test object, the required resolution and the application. The CMD revealed nearly all of the investigated defects as well or better than the EMD. Only for the scratch in the washing machine door glass (defect in region 1), the EMD achieved a slightly higher CNR than the CMD. In most cases the computation time of the CMD has been noticeably below that of the EMD. Regarding the performed experiments, the CMD outperforms the EMD with respect to the achieved contrast and the computation time and thus represents a sensible choice for the distance function $d\{\cdot,\cdot\}$.

In the sense of the criteria stated in Section 1.1.2, the presented methods are evaluated as follows:

1. **Versatility:** Based on the deflection maps captured by all three acquisition setups ($4f$-light field camera, schlieren deflectometer, laser deflection scanner) both the EMD and the CMD are capable of making opaque contaminants and scattering material defects visible.

2. **Accuracy:** For all performed experiments, the results of the EMD and the CMD meet the Rose criterion and the achieved CNRs are an order of magnitude higher than for the conventional method.

3. **Speed:** An inspection system based on a realization of the $4f$-light field camera would require about 10.5 s for the inspection of one test object. For the schlieren deflectometer the inspection time would be approximately 53 s and with the laser deflection scanner the inspection of one test object would require about 8 s.

4. **Adaptability:** As long as the whole test object can be imaged by the employed acquisition device, no special adaptation is required.

5. **Robustness:** Variations of the test object's geometry or its placement in the measurement field alter the captured deflection distributions. Due to the calculation of the gradient, these variations are irrelevant. Hence, inspection systems based on the processing of deflection maps as presented in this section are robust against such effects.

Compared to state-of-the-art systems for the visual inspection of transparent test objects with a complex geometry, the presented methods are superior with respect to all criteria but the required inspection time which is slightly longer. However, further revisions of the presented prototypes might achieve a shorter acquisition time. More efficient implementations of fastCMD in C++ instead of C# might yield an additional speed up.

6.2 PSD-signal processing

A laser deflection scanner equipped with a position sensitive detector (PSD) also yields the mean deflection direction $\mathbf{w}(\mathbf{m}) = (w_x(\mathbf{m}), w_y(\mathbf{m}))^\intercal$ for every measurement position $\mathbf{m}$. Material defects that lead to distant peaks in the deflection distribution of adjacent object points (cf. Fig. 1.5, transition from ray 4 to ray 5) also lead to large differences regarding their mean deflection direction. These differences can be spatially determined by first computing the Jacobian $\mathbf{J}_\mathbf{w}(\mathbf{m})$ of $\mathbf{w}(\mathbf{m})$ and then calculating its Frobenius norm $\|\mathbf{J}_\mathbf{w}(\mathbf{m})\|_\mathrm{F}$. The Jacobian $\mathbf{J}_\mathbf{w}(\mathbf{m})$ contains the partial derivatives of the mean deflection direction $\mathbf{w}(\mathbf{m})$ and its Frobenius norm $\|\mathbf{J}_\mathbf{w}(\mathbf{m})\|_\mathrm{F}$ yields scalar values that indicate the mentioned type of defect.

The Jacobian $\mathbf{J}_\mathbf{w}(\mathbf{m})$ can be estimated by applying an edge detection operator, e.g., the Sobel operator [Bey15], in m- and n-direction to $w_x(\mathbf{m})$ and $w_y(\mathbf{m})$:

$$\mathbf{J}_\mathbf{w}(\mathbf{m}) = \begin{pmatrix} \frac{\partial w_x(\mathbf{m})}{\partial m} & \frac{\partial w_x(\mathbf{m})}{\partial n} \\ \frac{\partial w_y(\mathbf{m})}{\partial m} & \frac{\partial w_y(\mathbf{m})}{\partial n} \end{pmatrix} \tag{6.21}$$

$$\approx \begin{pmatrix} \mathbf{S}_m ** w_x(\mathbf{m}) & \mathbf{S}_n ** w_x(\mathbf{m}) \\ \mathbf{S}_m ** w_y(\mathbf{m}) & \mathbf{S}_n ** w_y(\mathbf{m}) \end{pmatrix}, \tag{6.22}$$

with $**$ denoting the two-dimensional convolution operator.

The filter kernels $\mathbf{S}_m$, $\mathbf{S}_n$ of the Sobel operator are defined as:

$$\mathbf{S}_m := \begin{pmatrix} 1 & 0 & -1 \\ 2 & 0 & -2 \\ 1 & 0 & -1 \end{pmatrix}, \quad \mathbf{S}_n := \begin{pmatrix} -1 & -2 & -1 \\ 0 & 0 & 0 \\ 1 & 2 & 1 \end{pmatrix}. \tag{6.23}$$

The Frobenius norm of $\mathbf{J}_\mathbf{w}(\mathbf{m})$ is calculated via

$$\|\mathbf{J}_\mathbf{w}(\mathbf{m})\|_\mathrm{F} = \left(\left| \frac{\partial w_x(\mathbf{m})}{\partial m} \right|^2 + \left| \frac{\partial w_x(\mathbf{m})}{\partial n} \right|^2 + \right. \tag{6.24}$$

$$\left. \left| \frac{\partial w_y(\mathbf{m})}{\partial m} \right|^2 + \left| \frac{\partial w_y(\mathbf{m})}{\partial n} \right|^2 \right)^{\frac{1}{2}} \tag{6.25}$$

and can be further processed as a scalar image.

This approach will not be sensitive to absorbing defects which are only manifested by a reduced local intensity and not by scattering effects. This is because only the derivatives of the mean deflection direction is considered and not the absolute intensity. Furthermore, scattering defects that lead to a broad distribution of deflection directions (cf. Fig. 1.5, ray 3) will not result in a high Frobenius norm. The PSD-signal corresponds to the center of gravity of the deflection distribution of the incident light. If the intensity pattern is symmetric around the deflection direction corresponding to the defect-free case, the PSD-signal will be equal to that of the defect-free case and the defect cannot be revealed. This applies especially to small defects because they can cause diffraction effects which lead to symmetric intensity distributions.

6.2.1 Experiments

By means of the developed prototype of the laser deflection scanner, the PSD-signals have been captured for the automotive headlamp cover that has already been used as a test object in the experiments for the processing of deflection maps (see Section 6.1.3.2). Based on the PSD-signals, the Jacobain $\mathbf{J_w}(\mathbf{m})$ has been estimated and its Frobenius norm $\|\mathbf{J_w}(\mathbf{m})\|_F$ has been calculated. Figure 6.8 shows the resulting inspection image and also that one obtained with the CMD (see Section 6.1.2) for comparison. The following table lists the CNRs calculated according to Eq. (6.19) for the three numbered regions:

	CNR for region		
Method	1	2	3
CMD	**23.04**	**31.11**	**40.91**
PSD	17.92	24.95	10.19
Conventional bright field	4.29	22.86	4.29

(The highest CNR of every column is set in bold.)

Approximately the lower third of the inspection image resulting from processing the PSD-signals is affected by artifacts (magenta region) caused by the limitations of the prototypical implementation of the PSD in the laser deflection scanner. Because of that, defects located in that region cannot be revealed. By employing a PSD with a larger sensor area and by improving the precision of the positioning of the device, the artifacts can be avoided. Some of the defects, e.g., the upper left scattering defect highlighted by the blue rectangle and the absorbing defect visible in the magnified region 2 are visible in the image of the Frobenius norm but with a lower contrast compared to the CMD, as it is also stated by the computed CNRs. The remaining scattering defects in the magnified region 1 and the absorbing defect of the magnified region 3 are only barely revealed by processing the PSD-signals whereas they are shown with high contrast by the gradient calculation based on the CMD. This is because the corresponding deflection distributions have a broad intensity profile and cannot be distinguished from a distinct peak by means of the PSD-signals. Although the CMD performs best, the PSD still achieves higher CNRs compared to the conventional approach and allows to detect some of the material defects.

The PSD-signals of this test object could be acquired within only 1.2 s which is noticeably faster than capturing deflection maps. The mean time required for computing the Jacobian and the Frobenius norm has been 0.054 s $\pm$ 0.0013 s.

6.2.2 Discussion

In the following, the suitability of PSD-signal processing as a visual inspection system for complex-shaped transparent objects is discussed with respect to the criteria introduced in Section 1.1.2:

1. **Versatility:** Absorbing defects that also result in scattering of incident light as well as scattering defects leading to a shifted peak in the deflection distribution can be revealed. Defects which just broaden the deflection distribution cannot be made visible.

2. **Accuracy:** For the revealed defects, the achieved CNRs all satisfy the Rose criterion and are higher than those of the conventional bright field image. Hence, a reliable detection of corresponding defects is possible.

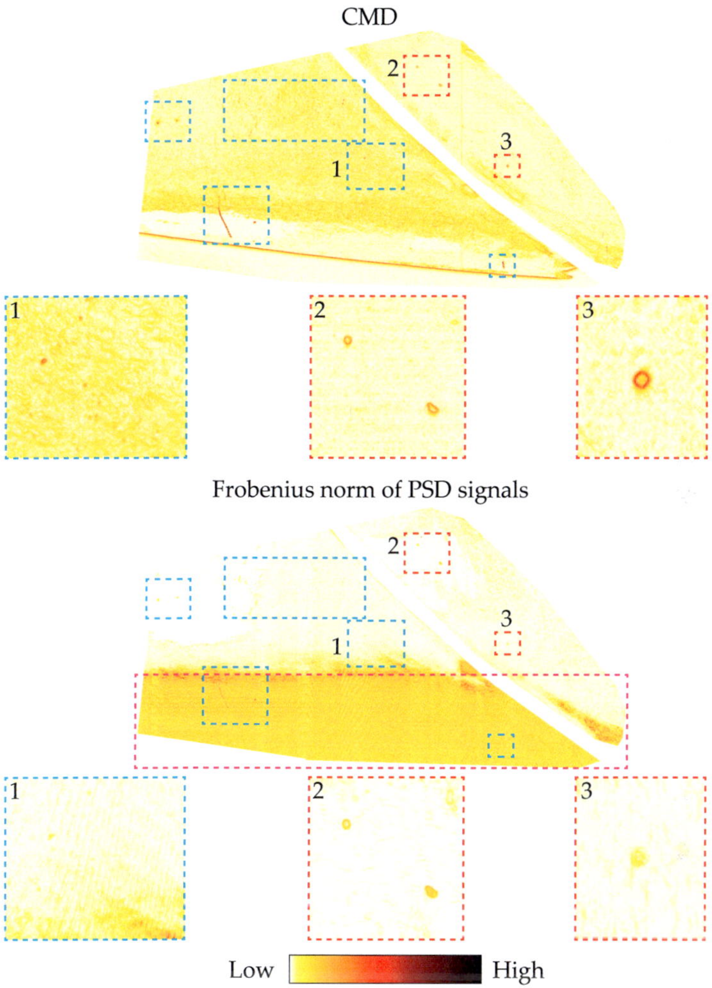

Figure 6.8: Inspection images of the investigated headlamp cover shown in pseudo colors. Intense edges and design structures of the test object have been masked out. Some of the defects are magnified under the respective inspection image.

3. **Speed:** Acquiring and processing the PSD-signals for a transparent object requires approximately 1.26 s.

4. **Adaptability:** As long as the whole test object can be imaged by the laser deflection scanner, no special adaptation is required.

5. **Robustness:** Variations of the test object's geometry or its placement in the measurement field alter the deflection distribution and hence the PSD-signals. Due to the calculation of the Jacobian, these variations are mitigated and the inspection system is robust against such effects.

In the case that the inspection time of the methods based on the processing of deflection maps is too high, acquiring and processing PSD-signals can represent a suitable alternative. The required inspection time of 1.26 s is superior to the state-of-the-art and many types of defects can be revealed.

6.3 Signal-to-noise ratio based image fusion

The Purity inspection system presented in the introduction (see Section 1.1.3) is also capable of inspecting such transparent test objects with a complex geometry. However, especially the determination of a suitable arrangement of the dark field light sources is a tedious task. By means of time-consuming experiments, experts have to design an optimal optical setup that renders material defects visible without producing disturbing reflections from intended design structures of the test object. This section introduces a visual inspection method based on image fusion algorithms which optimize an adequately formulated signal-to-noise ratio by exploiting information about the test object and the inspection system. The required information is automatically learned from images of defect-free objects and objects with artificially introduced defects. In the resulting inspection images, scattering defects are made visible with high contrast and intended object structures are suppressed. This reduces both the time and expertise needed for the system design. The images are assumed to be obtained with an optical setup similar to that of the Purity system, i.e., a combination of a bright field light source and several dark field light sources positioned around the test object. Images are captured while the test object is sequentially illuminated by one light source at a time resulting in a light field containing images of several different illumination modalities. In contrast to the Purity system, the dark field light sources require no optimal placement, together they just have to illuminate the whole volume of the test object. Since the inspection regarding opaque contaminants is already solved well by the Purity system and other approaches, the presented methods deal with revealing scattering material defects by means of the dark field light sources.

The following Section 6.3.1 describes the image formation procedure and motivates the image fusion approach. Section 6.3.2 explains how the required information about the test object can be obtained. Two image fusion strategies are introduced in Section 6.3.3. Section 6.3.4 describes the experiments performed to evaluate the presented methods and Section 6.3.5 closes the topic with a discussion of the inspection approach.

6.3.1 Image formation

Each of the C light sources illuminates the test object with a light field $L_c, c \in [1, \ldots, C]$. The test object transforms the incident light field into a modulated light field L'_c of which a camera image $g_c(\mathbf{x})$, i.e., a two-dimensional projection, is captured by the telecentric camera (cf. optical setup of the Purity system shown in Fig. 1.4). The single images corresponding to the different light sources can be summarized in a C-channel image

$$\mathbf{g}(\mathbf{x}) = (g_1(\mathbf{x}), g_2(\mathbf{x}), \ldots, g_c(\mathbf{x}), \ldots, g_C(\mathbf{x}))^\mathsf{T} . \qquad (6.26)$$

Depending on the position and orientation of a light source with respect to the test object and the camera, defects in some regions of the test object will be revealed, whereas the system will be blind for other parts of the object as illustrated by Fig. 6.9. Some of the emitted light rays will cause direct reflections at geometric structures or the surface of the test object (dotted optical paths in Fig. 6.9). If the reflected rays reach the camera, the resulting camera image will show bright pixels at the corresponding positions for defect-free objects as well as for objects affected by defects at these positions. The inspection system would be blind at the object positions of the direct reflections concerning this light source and defects located at these positions would not be found (blue light source in Fig. 6.9a). For another light source with a different location and orientation, the positions of the reflections in the camera image would be different but the defect would still scatter some of the incident light towards the camera so that it would be revealed (green light source in Fig. 6.9b). If the positions of the reflections are known for each light source, they can be omitted in an image fusion process resulting in an image free from reflections. If additional information about the spatially resolved capability of revealing defects is available for every light source, a weighting could be applied during the fusion process. Since the geometry of the test objects can be affected by small variations caused by their production process, stochastic variations have to be taken into account.

In order to formulate image fusion algorithms that are able to extract the defects from $\mathbf{g}(\mathbf{x})$ while suppressing the reflections, the following approach is pursued. The image formation process is mod-

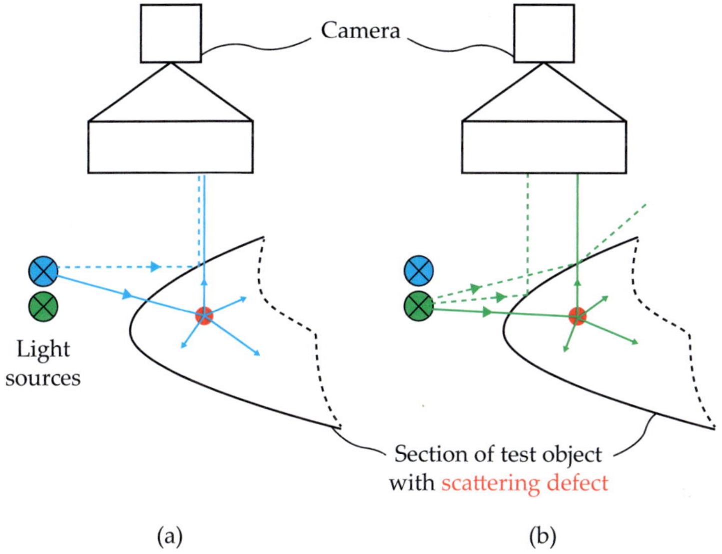

Figure 6.9: Influence of the geometry of the test object on the image formation process. The test object, which is affected by a scattering defect, is illuminated by two light sources (blue and green) and is observed by a telecentric camera. (a) The reflection of the light rays of the blue light source at the test object's surface are imaged at the same image position as the light rays scattered by the defect, rendering the defect invisible. (b) The reflected light rays of the green light source miss the camera or are imaged at another image position as the light rays scattered by the defect which is why the defect is visible.

eled as a summation of a wanted signal component $\mathbf{s}(\mathbf{x})$ and an unwanted noise component $\mathbf{n}(\mathbf{x})$:

$$\mathbf{g}(\mathbf{x}) = \mathbf{s}(\mathbf{x}) + \mathbf{n}(\mathbf{x})\,. \tag{6.27}$$

Image regions showing the defects represent the signal component and all other image structures, e.g., reflections caused by the surface of the object or by its intended structures, represent the noise component. In order to enhance the signal quality for a noise-affected image formation process as shown above, a well-known strategy is to maximize the *signal-to-noise ratio (SNR)* [Gon08]. In the field of classical signal processing, the SNR is usually defined as the quotient

$$\mathrm{SNR} = \frac{P^{\mathrm{signal}}}{P^{\mathrm{noise}}} \tag{6.28}$$

of the power P^{signal} of the wanted signal component and the power P^{noise} of the (unwanted) noise component. The SNR can be increased by emphasizing the signal component and by reducing the influence of the noise component.

Information about possible material defects can be utilized to increase the signal component. Unfortunately, the set of material defects that might occur is usually unknown a-priori. The factors influencing the noise component are closely linked to the properties of the test object and the optical components of the inspection system. Information about the test objects and the inspection system, e.g., the positions of the unwanted reflections caused by each light source, can provide the necessary information for improving the SNR by mitigating the noise component. Some information about the signal component, i.e., the appearance of the defects, may also be acquired. For this purpose, artificial defects can be incorporated into a test object at different positions in order to spatially measure their response with respect to different illumination sources.

Since the manual incorporation of such information can be difficult and time-consuming, a method will be presented which automatically acquires this information based on defect-free test object samples.

The presented inspection approach is divided into two stages. In the first step, which is explained in Section 6.3.2, information about the test object and the inspection system is acquired. The image fusion algorithms, described in Section 6.3.3, constitute the second step and use the learned information together with a captured multichannel input image of the test object. In the generated output image, defects are made visible with high contrast whereas disturbing reflections are suppressed.

6.3.2 Acquisition of exploitable information

In order to be able to improve the SNR of the acquired images, exploitable information about the contained signal component and noise component have to be obtained. Since the test object is affected by variations caused by the manufacturing process, both the signal component $s(\mathbf{x})$ and the noise component $\mathbf{n}(\mathbf{x})$ have to be considered as random vectors:

$$\mathsf{g}(\mathbf{x}) = \mathsf{s}(\mathbf{x}) + \mathbf{n}(\mathbf{x}) \,. \tag{6.29}$$

Information about the signal component $\mathsf{s}(\mathbf{x})$ regards the capability of the system's channels to reveal material defects. It depends on the position $\mathbf{x}$, as the single illumination sources do not achieve a homogeneous illumination of the whole test object volume in general. As the noise component contains the influence of the local surface geometry of the test object, it is also a function of the position $\mathbf{x}$. In order to be able to capture the position-dependent sensitivity and noise component, the image acquisition setup is required to be spatially fixed, i.e., in the captured images, the test objects always need to have the same position and orientation.

6.3.2.1 Noise component

The noise component is influenced by many different factors of the test object's manufacturing process. By assuming that all these factors are independent and identically distributed, $\mathbf{n}(\mathbf{x})$ can be supposed to be normally distributed according to the central limit theorem [Pap02]. Hence, the mean characteristics of the noise component can be characterized via the expected values $\boldsymbol{\mu}(\mathbf{x})$ for each position $\mathbf{x}$ and each channel. These values can be obtained by capturing images of a preferably large set of defect-free test objects. For a completely defect-free test object, it holds,

$$\mathbf{s}(\mathbf{x}) \equiv 0\,, \tag{6.30}$$

and thus

$$\mathbf{g}(\mathbf{x}) = \mathbf{n}(\mathbf{x})\,. \tag{6.31}$$

It is usually impossible to obtain a set of test objects of which all are completely defect-free (and free from dust particles, which can also cause high signal values in dark field setups). To overcome this issue, outlier detection methods can be employed or a large set of test objects can be used. For such a training set $\mathcal{T}_{\mathrm{n}}$ of defect-free objects, the expected values are estimated via

$$\hat{\mu}_c(\mathbf{x}) = \frac{1}{|\mathcal{T}_{\mathrm{n}}|} \sum_{\mathbf{g} \in \mathcal{T}_{\mathrm{n}}} g_c(\mathbf{x})\,, \quad c = 1, \ldots, C\,. \tag{6.32}$$

The obtained $\hat{\boldsymbol{\mu}}(\mathbf{x}) = (\hat{\mu}_1(\mathbf{x}), \ldots, \hat{\mu}_C(\mathbf{x}))^{\mathsf{T}}$ can now be used to judge the influence of the noise component to an observed signal and hence to obtain an optimized inspection image as it is shown in Section 6.3.3.

6.3.2.2 Signal component

As stated previously, the inspection system's sensitivity for defects can vary between positions $\mathbf{x}$ and between channels c. To assess the sensitivity characteristics, a specially designed test object is required. In this, so-called *signal sample*, artificial defects are introduced, e.g., small air bubbles can be created inside the test object volume via laser engravings. As it is not feasible to produce many of such signal samples, the stochastic variations of $\mathbf{s}(\mathbf{x})$ cannot be determined by means of calculating statistics over a series of test objects. Instead, a spatial averaging is performed which mitigates the influence of image noise. By acquiring a c-channel inspection image of the signal sample, an estimation $\hat{\mathrm{s}}(\mathbf{x})$ of the signal component can now be obtained at the support positions $\mathbf{x} \in \mathcal{S}$ of the artificial defects and interpolated in between. Based on $\hat{\mathrm{s}}(\mathbf{x})$, the signal power and the SNR can be estimated as it will be shown in Section 6.3.3.2. The defects should be arranged on a preferably dense Cartesian grid to ensure an accurate estimation of $\hat{\mathrm{s}}(\mathbf{x})$.

6.3.3 Image fusion strategies

Based on the acquired information about the signal component $\mathbf{s}(\mathbf{x})$ and the noise component $\mathbf{n}(\mathbf{x})$, an inspection image that increases the SNR specified in Eq. (6.28) is to be generated. Depending on how much information about the two components is available, different fusion strategies can be applied.

6.3.3.1 Noise-based selective channel fusion

If no signal sample is available, exploitable information can only be acquired regarding the noise component. In such a case, a fusion strategy has to perform the image fusion using only the knowledge about the noise component. A simple but effective approach is to determine the image regions $\mathbf{v}(\mathbf{x})$, which are free from disturbing reflections for each channel of the imaging system, i.e., for which the estimated mean noise levels $\hat{\boldsymbol{\mu}}(\mathbf{x})$ are below a threshold t:

$$v_c(\mathbf{x}) = \begin{cases} 1, & \text{if } \hat{\mu}_c(\mathbf{x}) < t \\ 0, & \text{otherwise}. \end{cases} \tag{6.33}$$

The indicator function $v_c(\mathbf{x})$ determines for each channel c and each position $\mathbf{x}$ whether the respective information will be used ($v_c(\mathbf{x}) = 1$) or discarded ($v_c(\mathbf{x}) = 0$) during the fusion process.

The result of the image fusion for an input image $\mathbf{g}(\mathbf{x})$ is obtained via

$$g'(\mathbf{x}) = \frac{1}{\sum_{c=1}^{C} v_c(\mathbf{x})} \sum_{c=1}^{C} g_c(\mathbf{x}) \cdot v_c(\mathbf{x}). \tag{6.34}$$

For each pixel $\mathbf{x}$, $g'(\mathbf{x})$ is the average of the gray values of those channels, which are not affected by reflections at the respective position.

The information contained in the contributing channels are all weighted equally, i.e., they are all considered to be equally meaningful. However, a possibly present material defect might not lead to increased signal values for all illumination directions. Channels not showing the defect can lead to a reduced signal level in the resulting $g'(\mathbf{x})$ and therefore impede the detection of the defect by further processing steps.

6.3.3.2 SNR-based weighted channel fusion

If a signal sample is available as described in Section 6.3.2.2, information about the signal component can be acquired in addition to the noise component in order to improve the image fusion process.

The artificially introduced defects cover several pixels in the image of the signal sample. The signal power $P_c^{\mathrm{signal}}(\mathbf{x})$ of channel c can be estimated by computing the sum

$$P_c^{\mathrm{signal}}(\mathbf{x}) = \frac{1}{|\mathcal{U}_{\mathrm{s}}(\mathbf{x})|} \sum_{\mathbf{x}' \in \mathcal{U}_{\mathrm{s}}(\mathbf{x})} \hat{s}_c(\mathbf{x}') \tag{6.35}$$

of the pixel values of $\hat{s}(\mathbf{x})$ inside a region $\mathcal{U}_{\mathrm{s}}(\mathbf{x})$ defining a neighborhood around each support position $\mathbf{x} \in \mathcal{S}$. Similarly, an estimation of the noise power $P_c^{\mathrm{noise}}(\mathbf{x})$ of channel c can be obtained:

$$P_c^{\mathrm{noise}}(\mathbf{x}) = \frac{1}{|\mathcal{U}_{\mathrm{n}}(\mathbf{x})|} \sum_{\mathbf{x}' \in \mathcal{U}_{\mathrm{n}}(\mathbf{x})} \hat{\mu}_c(\mathbf{x}') \,. \tag{6.36}$$

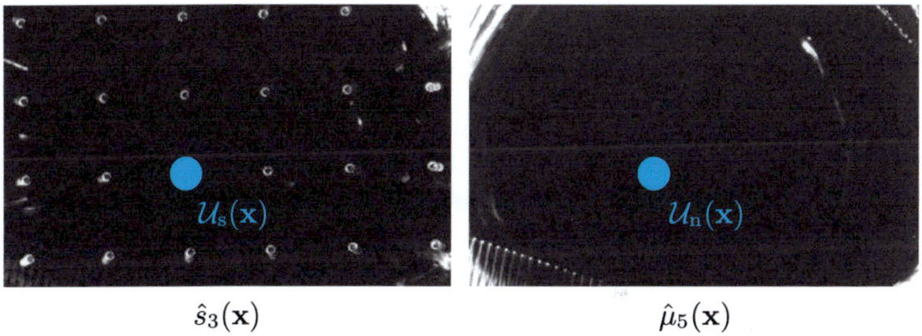

Figure 6.10: Example for the estimation of the power of the signal and the noise component. The left image shows a section of the 3rd channel of the estimated signal component $\hat{s}_3(\mathbf{x})$ and the right image shows the 5th channel of the estimated noise component $\hat{\mu}_5(\mathbf{x})$. The blue areas mark the positions of the pixels considered for the power estimation which are contained in $\mathcal{U}_{\mathrm{s}}(\mathbf{x})$ and $\mathcal{U}_{\mathrm{n}}(\mathbf{x})$, respectively.

For each position $\mathbf{x}$, the functions $\mathcal{U}_{\mathrm{s}}(\mathbf{x})$, respectively, $\mathcal{U}_{\mathrm{n}}(\mathbf{x})$ provide sets of coordinates representing circular areas that cover an artificial defect's image region, respectively, the region that is employed to estimate the noise power (see Fig. 6.10 for an illustration). By this means, the signal-to-noise ratio $\mathrm{SNR}_c(\mathbf{x})$ can now be calculated for each channel c and for each support position $\mathbf{x} \in \mathcal{S}$:

$$\mathrm{SNR}_c(\mathbf{x}) = \frac{P_c^{\mathrm{signal}}(\mathbf{x})}{P_c^{\mathrm{noise}}(\mathbf{x})}, \quad \mathbf{x} \in \mathcal{S}. \tag{6.37}$$

To estimate the SNR-values for positions $\mathbf{x} \notin \mathcal{S}$, an interpolation $\mathrm{Interpol}\{\cdot\}$-operator (e.g., a bilinear interpolation) has to be applied to obtain an interpolation $\widehat{\mathbf{SNR}}(\mathbf{x})$ of $\mathbf{SNR}(\mathbf{x})$:

$$\widehat{\mathbf{SNR}}(\mathbf{x}) = \mathrm{Interpol}\{\mathbf{SNR}(\mathbf{x})\}. \tag{6.38}$$

Accordingly, the SNR-based weighted channel fusion can now be applied to a c-channel input image $\mathbf{g}(\mathbf{x})$:

$$g'(\mathbf{x}) = \frac{1}{\sum_{c=1}^{C} \widehat{\mathrm{SNR}}_c(\mathbf{x})\, v_c(\mathbf{x})} \sum_{c=1}^{C} g_c(\mathbf{x})\, \widehat{\mathrm{SNR}}_c(\mathbf{x})\, v_c(\mathbf{x}). \tag{6.39}$$

During the fusion process, the pixels of the input image $\mathbf{g}(\mathbf{x})$, which are not affected by reflections, i.e., for which $\mathbf{v}(\mathbf{x}) = 1$, are weighted with the SNR of the respective channel c and position $\mathbf{x}$. So, the sensitivity of the imaging system for possibly present material defects is taken into account to improve the fusion result.

6.3.4 Experiments

Several experiments have been performed to evaluate the described approaches. Images have been acquired with the Purity inspection setup as shown in Fig. 1.4. A telecentric camera system is focused on the test object. Several light sources are arranged in a dark field constellation. Transparent plastic lenses with a complex geometry and strong surface shape features were used as the test objects. The setup has been equipped with $C = 20$ different dark field light sources arranged around the test object. An image is captured for each light source turned on separately, resulting in 20-channel images for each test object. The whole acquisition took approximately 2 s. Figure 6.11 shows sample images of four channels. Each light source causes characteristic reflections at the test object's surface and at the ripple-structured shape features making the respective positions blind for defects in the corresponding channels. As described in Section 6.3.2.1, the image signals caused only by the reflections and not by material defects are considered as the noise component. By acquiring the set $\mathcal{T}_{\mathrm{n}}, |\mathcal{T}_{\mathrm{n}}| = 3$ of image-series of defect-free test objects, the mean influence of the noise component $\hat{\mu}(\mathbf{x})$ is estimated according to Eq. (6.32). Thresholding the single channels of $\hat{\mu}(\mathbf{x})$ using Eq. (6.33) yields the indicator function $\mathbf{v}(\mathbf{x})$ defining the regions that can be considered for the image fusion. With the help of the images of a signal sample (cf. Section 6.3.2.2), an estimation $\hat{\mathrm{s}}(\mathbf{x})$ of the signal component is obtained. Based on $\hat{\mathrm{s}}(\mathbf{x})$ and $\hat{\mu}(\mathbf{x})$, the $\mathbf{SNR}(\mathbf{x})$ is estimated at the support positions $\mathbf{x} \in \mathcal{S}$ of the signal sample and interpolated to obtain $\widehat{\mathbf{SNR}}(\mathbf{x})$.

Using the acquired information, the two fusion strategies described in Section 6.3.3 were applied to three test objects affected by different material defects. One application of one of the fusion algorithms required $0.664 \text{ s} \pm 0.007 \text{ s}$.[2] The following visualizations are shown in pseudo colors. Figures 6.12, 6.13 and 6.14 show example inspection images for three test objects. They all show a center section of the test object, so that the material defects are visible to the reader. The upper row shows inspection images as they would be acquired using a conventional inspection approach. They allow comparing the results of the fusion strategies. The upper left-hand image corresponds to

[2] Technical details of the employed computer system are provided in Appendix A.3.

a bright field setup using an area light source behind the test object and the upper right-hand image is a dark field capture where all the 20 illumination sources have been turned on simultaneously and no fusion strategy has been employed.

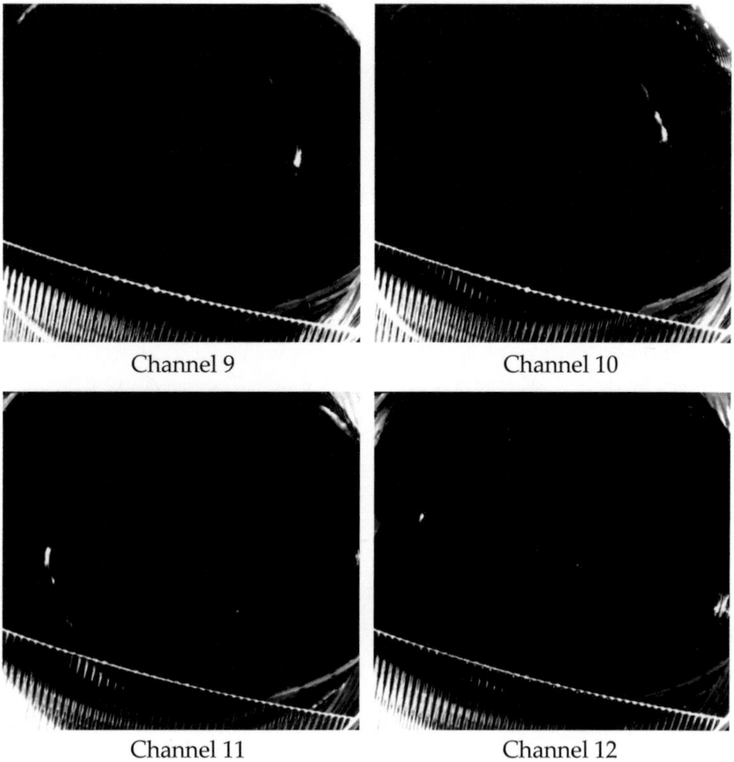

Figure 6.11: Examples of single-channel camera images of a defect-free test object. The images clearly show the strong reflections that are caused at different positions depending on the corresponding light source.

The second row shows the images resulting from the noise-based selective channel fusion strategy described in Section 6.3.3.1 (left-hand side) and the SNR-based weighted channel fusion strategy introduced in Section 6.3.3.2 (right-hand side). A discussion of the details of the inspection results and the differences that can be observed between the strategies follows.

6.3.4.1 Test object 1

The inspected test object is affected by three major material defects (see Fig. 6.12, blue and white regions). The following table lists the CNRs calculated according to Eq. (6.19) for the corresponding regions:

	CNR for region		
Method	1	2	3
Bright field	23.45	10.12	21.03
All channel dark field	68.24	110.59	120.59
Noise-based selective fusion	61.33	96.89	222.44
SNR-based weighted fusion	**282.50**	**215.00**	**316.25**

(The highest CNR of every column is set in bold.)

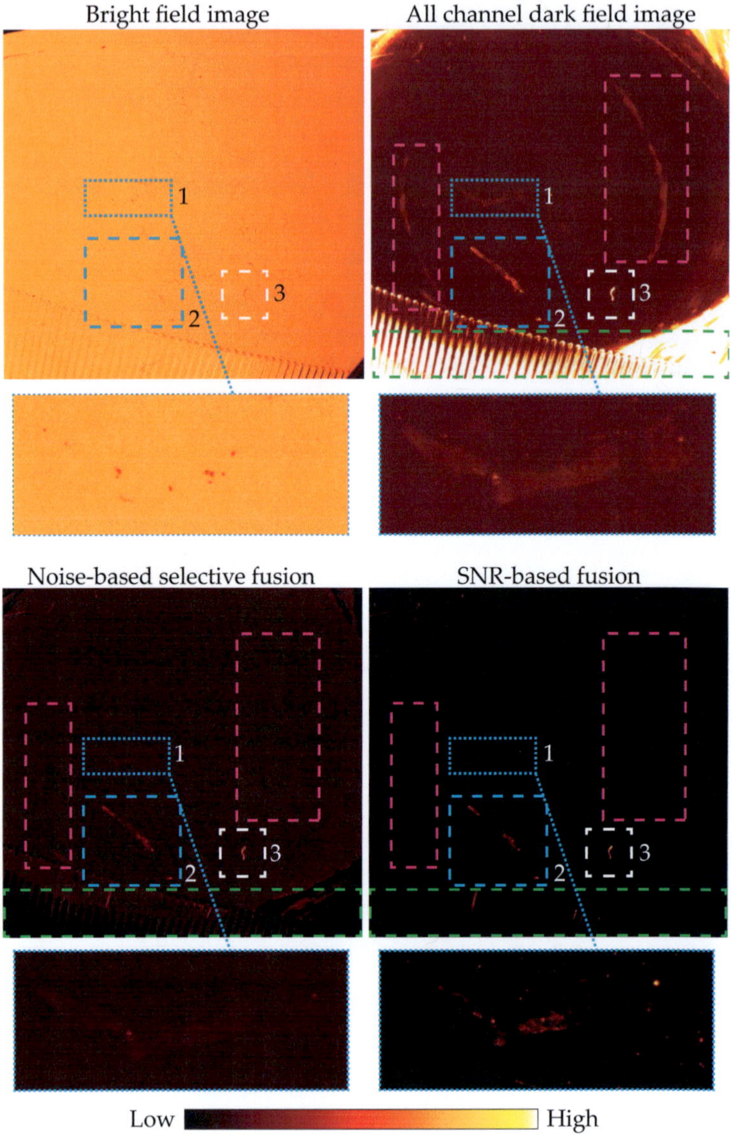

Figure 6.12: Inspection images for test object 1 shown in pseudo colors.

In the dark field image using all channels simultaneously the defects are visible, however, there are also strong reflections caused either by direct surface reflections of the light sources into the camera (magenta regions) or by light scattering of shape features of the test object (green region). These reflections have similar intensity values as the defects' images making it hard or even impossible to distinguish between defects and reflections based on the intensity levels. In the images resulting from the fusion strategies, the reflections have been suppressed nearly completely. The reflections in the right part of the green region are so severe that the fusion strategies nearly completely discard the respective information, which is the reason why the corresponding regions are almost entirely black in the fused images.

The defect located in the white region (defect 3) most likely is an opaque and absorbing defect, which is why it is also made visible by the bright field illumination, although only with low contrast. Two scratches are present inside the blue regions (defects 1 and 2). Due to their scattering nature, these defects are revealed best by the dark field illumination and the two fusion strategies. The defect marked by the dotted blue region is rather weak, which is why it is only slightly visible with the noise-selective channel fusion strategy. The second fusion approach also clearly shows this defect. For defects 1 and 2, the conventional dark field image results in higher CNRs than the noise-based fusion strategy. This is because the defects are not located near one of the surface reflections what would reduce the corresponding CNR. The SNR-based weighted channel fusion strategy yields the highest CNRs for all three defects as expected.

6.3.4.2 Test object 2

The second test object shows multiple material defects (see Fig. 6.13). The CNRs calculated according to Eq. (6.19) for the corresponding regions are listed in the following table:

	CNR for region		
Method	1	2	3
Bright field	8.91	14.04	12.26
All channel dark field	38.43	97.38	1.28
Noise-based selective fusion	221.78	118.33	12.67
SNR-based weighted fusion	**419.67**	**165.36**	**165.23**

(The highest CNR of every column is set in bold.)

Some of the defects are opaque contaminants and are visible in all inspection images (white regions). Parts of the remaining, scattering defects are located in the ripple-structured area (blue dashed regions). In the conventional dark field setup, this zone is dominated by reflections caused by the object's shape features, which is why these defects are barely visible in the respective image. In contrast, the images resulting from the two fusion strategies clearly show the defects as it is also supported by the computed CNRs. Some small scattering defects are made visible best by the SNR-based weighted channel fusion strategy (dotted blue region, defect 2). In the respective region, the estimated $\widehat{\mathbf{SNR}}(\mathbf{x})$ is comparatively high, resulting in a high amplification of the corresponding pixels during the fusion process. As for the first test object, the image acquired with the conventional dark field setup is affected by reflections (magenta and green regions) which are suppressed by the image fusion methods.

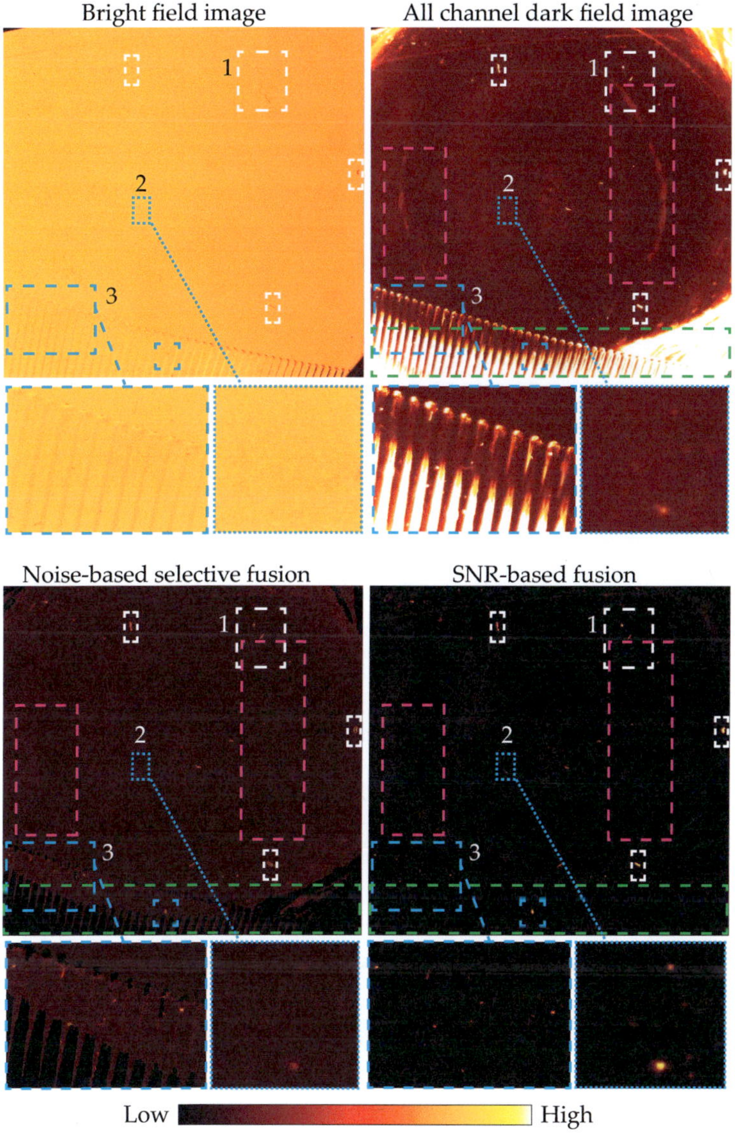

Figure 6.13: Inspection images for test object 2 shown in pseudo colors.

6.3.4.3 Test object 3

Figure 6.14 shows sections of the inspection images of test object 3. The bright field image shows several absorbing contaminants in or on the test object. In the upper right-hand corner of the shown section (blue region), there are multiple small air bubbles which are enclosed inside the test object. The following table lists the CNRs calculated according to Eq. (6.19) for this region:

Method	CNR for region 1
Bright field	5.16
All channel dark field	29.28
Noise-based selective fusion	17.88
SNR-based weighted fusion	**171.29**

(The highest CNR is set in bold.)

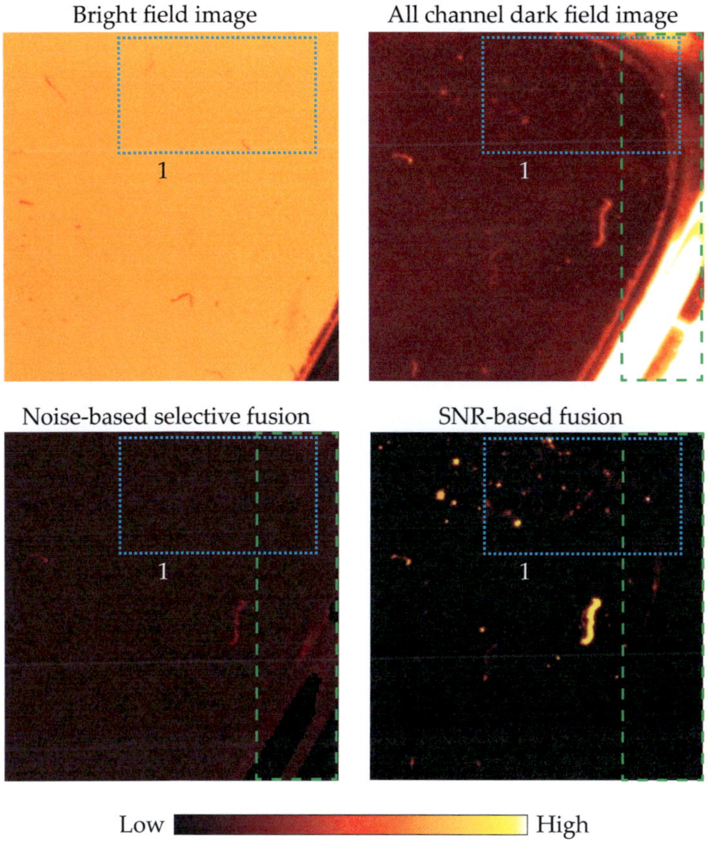

Figure 6.14: Magnified section of the inspection images for test object 3 shown in pseudo colors.

The absorbing structures are made visible by the bright field image but are not shown by the other inspection images since they cause no scattering. The enclosed air bubbles are not revealed by the bright field image because they absorb nearly any light. In the conventional dark field, these defects are slightly visible, however, only with low contrast. Furthermore, once again strong reflections (green region) caused by the test object's shape features are present in the dark field image and compete with the visualization of the contaminants. The reflections are reduced by both fusion approaches. Nevertheless, the air bubbles are not shown in the image corresponding to the noise-based selective channel fusion. Most likely, only a few of the 20 dark field light sources illuminated the inclusions in a direction leading to scattering events observable by the camera. As the averaging performed by the noise-selective channel fusion strategy treats all noise-unaffected channels equally, the information about the inclusions was reduced to the background intensity level. In contrast, the SNR-based weighted channel fusion utilizes the additional information gained from the signal sample and is able to assign more weight to channels showing a more intense response to defects in the respective region of the test object. Hence, the small inclusions can be clearly seen in the inspection image resulting from the SNR-based weighted channel fusion strategy. As expected, the highest CNR is achieved by the SNR-based weighted channel fusion. The CNR of the conventional dark field image is higher than that of the noise-based selective channel fusion. This is because the defects are not made visible by all channels and hence get alleviated by the performed unweighted averaging.

6.3.5 Discussion

This section has introduced two image fusion strategies that exploit different amounts of information about the image formation process. By using this knowledge, it is possible to enhance the signal of defects and to reduce the influence of the noise, i.e., direct reflections of the illuminating light at the surface of the test objects. The proposed learning procedure is realized by an automated acquisition of image series of defect-free test objects and, if applicable, of a signal sample with artificially introduced defects. Even non-expert users are able to introduce new types of objects by this procedure.

The experiments show that, by appropriately modeling the signal and noise components and by acquiring suitable exploitable information, improved inspection results can be achieved using image fusion strategies. If only information about the noise component is available, the presented noise-based channel fusion can be applied. This already leads to improved results as the influence of surface reflections or scatterings caused by the test object are mitigated and material defects are made visible with high contrast. Since the method does not take the signal component (i.e., the system's sensitivity with respect to the channel and the position) into account, small material defects, which are only visible in few channels, might not be conserved by the fusion process.

When information about the signal component can be acquired (e.g., by means of a signal sample) the SNR can be locally estimated for each channel, which allows a more specific channel fusion. The SNR-based weighted channel fusion applies weights to the single channels according to the respective SNR. This approach also successfully mitigates the influence of unwanted reflections. Furthermore, due to its sensitivity to the SNR, it reveals even small transparent inclusions, like air bubbles, which are only present in a few channels of the acquired image (cf. results of test object 3, Fig. 6.12). This makes the SNR-based weighted channel fusion strategy superior to the purely noise-based approach. At positions where the signal component is very low compared to the noise component in all channels, the SNR-based fusion will also result in black and therefore uninspectable regions. This issue could be resolved by improving the illumination constellation, such as by adding further light sources that yield higher signal components at the respective positions.

With respect to the evaluation criteria introduced in Section 1.1.2, the two image fusion methods are evaluated as follows:

1. **Versatility:** Both of the presented algorithms are capable of revealing opaque and scattering material defects.

2. **Accuracy:** The algorithms achieve CNRs that are either considerably higher or approximately equal compared to those of the conventional approaches. Due to the suppression of disturbing reflections, the presented methods further increase the system's capability of detecting defects.

3. **Speed:** The inspection of one test object requires approximately 3 s.

4. **Adaptability:** The noise-based selective channel fusion can easily be adapted to new types of test objects since only a small set of defect-free test objects is required. For the SNR-based weighted channel fusion, a signal sample has to be manufactured what might involve several steps and a noticeable amount of time.

5. **Robustness:** Since information about the noise component is learned based on a set of test objects, small variations of the test object's geometry caused by the manufacturing process are tolerated by the presented methods. As described before, the response of the defects, i.e., the signal component, can be assumed to independent of small variations of the object's geometry. The introduced algorithms are very sensitive to misalignments of the test object as the signal and the noise component are spatially referenced.

6.4 Light transport matrix processing

In Section 4.5, the concept of light transport matrices and an optical setup for their acquisition with the goal of inspecting transparent test objects for scattering defects has been introduced. For a camera with $S = M \times N$ pixels and a programmable light source with $Q = U \times V$ pixels, the light transport can be characterized by the transport matrix

$$\mathbf{T} = \begin{pmatrix} \mathbf{c}(1,1)^\mathsf{T} \\ \mathbf{c}(1,2)^\mathsf{T} \\ \vdots \\ \mathbf{c}(M,N-1)^\mathsf{T} \\ \mathbf{c}(M,N)^\mathsf{T} \end{pmatrix}, \tag{6.40}$$

with the row vectors $\mathbf{c}(m,n)$ denoting so-called *correspondence vectors* [OTo10]. Each row of $\mathbf{T}$, i.e., each correspondence vector

$$\mathbf{c}(m,n) = \begin{pmatrix} c_{1,1}(m,n) \\ c_{1,2}(m,n) \\ \vdots \\ c_{U,V-1}(m,n) \\ c_{U,V}(m,n) \end{pmatrix} \tag{6.41}$$

holds the contributions of the $U \cdot V$ light source pixels to the signal captured by camera pixel $(m,n)^\mathsf{T}$.

6.4.1 Feature extraction

To reveal defects inside a transparent test object based on its light transport matrix, suitable features have to be defined. Therefore, Fig. 6.15 shows how scattering defects are assumed to be manifested in $\mathbf{T}$. The test object, a plano-convex lens, is affected by a scattering material defect. The telecentric camera's rays of sight passing defect-free regions of the test object are deflected due to the object's surface geometry and they finally reach a pixel of the programmable light source. Hence, the respective correspondence vectors show only a distinct peak at the element corresponding to the hit light source pixel. In contrast, the ray of sight which is incident to the material defect is scattered into multiple directions. The scattered rays of sight reach several light source pixels which all contribute to the signal of the respective camera pixel. Consequently, the intensities in the associated correspondence vector have a broader distribution. Based on these considerations, two features are introduced that can be extracted out of a light transport matrix and that are suitable for detecting scattering material defects.

Feature ScatterCount As discussed, scattering defects cause the light of multiple pixels of the programmable light source to be scattered into multiple directions. Hence, some of the scattered light rays also propagate parallel to the optical axis, are captured by the telecentric camera and contribute to the signal of the pixel onto which the defect is imaged (cf. the third example correspondence vector shown in Fig. 6.15). Consequently, many elements of the respective correspondence vector $\mathbf{c}(m,n)$ are greater than a threshold t. This motivates the feature *ScatterCount*

$$\mathrm{sc}(m,n) := |\{i \in [1,\ldots,UV] : c_i(m,n) \geq t\}| \tag{6.42}$$

that contains for each camera pixel $(m, n)^\intercal$ the count of components of $\mathbf{c}(m, n)$ which are greater than t. Therefore, the threshold t has to be adequately chosen with respect to the background noise level.

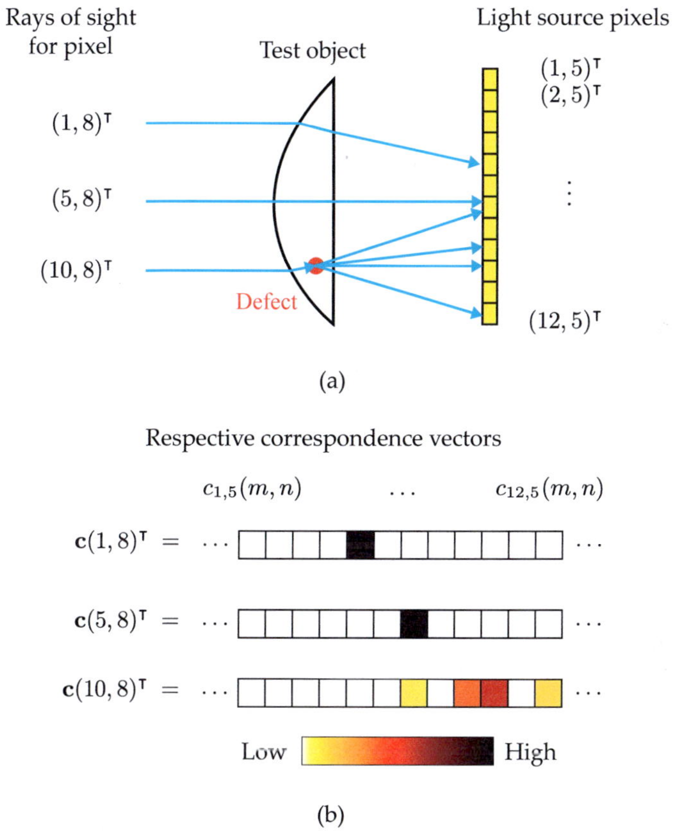

Figure 6.15: Illustration of the assumed manifestation of scattering material defects in the light transport matrix. (a) A plano-convex lens represents an example test object. The rays of sight corresponding to three pixels of the telecentric camera are shown in blue. Depending on their path through the test object, they hit different pixels of the programmable light source. In the case of hitting a scattering defect, the ray of sight is scattered and hits multiple light source pixels. (b) The intensity distributions of the respective correspondence vectors are shown in pseudo colors.

Feature ScatterWidth Not only the number of light source pixels contributing to the signal of a detector element can contain information for revealing a scattering material defect but also the maximum distance between those pixels on the light source. This distance is captured by the feature *ScatterWidth*, which is defined as:

$$\mathrm{sw}(m,n) := \begin{cases} 0, & \text{if} \max\limits_{i\in[1,\ldots,UV]} c_i(m,n) < t, \\ E - A, & \text{otherwise}, \end{cases} \tag{6.43}$$

$$A = \min\{i : c_i(m,n) \geq t\}, \tag{6.44}$$

$$E = \max\{i : c_i(m,n) \geq t\}. \tag{6.45}$$

Here, A is the index of the first and E the index of the last entry of $\mathbf{c}(m,n)$ that are greater than a threshold t which also has to be chosen with respect to the background noise level.

6.4.2 Experiments

To evaluate the introduced features for finding scattering defects based on the light transport matrix of transparent test objects, simulated experiments have been performed. By means of the rendering framework Mitsuba and the additional plugins introduced in Chapter 5, a virtual inspection setup as shown in Fig. 4.17 has been set up. The light transport matrix could be directly extracted out of the rendering process, so no illumination sequence was needed. The telecentric camera has been parametrized to have a spatial resolution of 100×100 and the employed programmable light source had 500×500 pixels. Since for simulating a light transport matrix the rendering framework has to simulate much more samples per camera pixel than for rendering a conventional image, comparatively low resolutions had to be chosen for the camera and the light source to keep the rendering time in the order of one day. A double-convex lens suited as the test object. Small air bubbles with cross-section areas of $r = 0.5, 1.0, 1.5$ times the projected area of a camera pixel have been introduced into the test object to simulate scattering material defects. For all three defect sizes, three instances of the air bubble with different positions have been created, starting from the test object's center towards its left border (with respect to the simulated images). Based on the acquired light transport matrix, the two introduced features have been calculated. For comparison, conventional inspection images of a telecentric camera with the same resolution used in combination with an area light source placed behind the test object have also been simulated.

Figure 6.16 shows the resulting inspection images and the following table lists the CNRs which have been computed for the defect in the center of the lens:

	CNR for defect with		
Method	$r = 0.5$	$r = 1.0$	$r = 1.5$
ScatterCount	**17.92**	**42.37**	**68.55**
ScatterWidth	13.82	16.55	19.45
Conventional bright field	9.62	25.77	35.00

(The highest CNR of every column is set in bold.)

The conventional image and the image corresponding to the feature ScatterCount clearly allow to identify the first test object (first row) as being intact, i.e., free from defects. The feature ScatterWidth shows many noise artifacts caused by the rendering process which could falsely be recognized as defects. This is why it achieves the lowest CNRs for all defect sizes. The defects of all sizes are shown with the highest contrast by the feature ScatterCount, as it is also stated by its CNRs. Although the defects are only slightly visible in the conventional inspection image, the corresponding CNRs are higher than those of the feature ScatterWidth for the two larger defect sizes. This is because the conventional inspection image has the lowest noise level and hence low image values at the positions of the defects are sufficient to yield a higher CNR. At the right-hand side of the test object, defect-like intensities can be seen in the inspection image corresponding to the feature ScatterWidth. Since there is no defect present at that position, the visible structure represents an artifact. Examining the light transport matrix at the respective positions showed that these artifacts are caused by very weak reflections inside the test object of the left-most defect. Based on these experiments, the feature ScatterCount represents a suitable approach for processing light transport matrices for the detection of scattering defects.

Computing the features of the light transport matrix took 17.9 s $\pm$ 0.2 s for the feature ScatterCount, 17.9 s $\pm$ 0.4 s for the feature ScatterWidth, respectively 18.1 s $\pm$ 0.5 s for computing both features simultaneously. Obviously, since extracting the two features simultaneously is not notably slower than extracting only one of them, the required computation time is mainly spent accessing the huge light transport matrix.

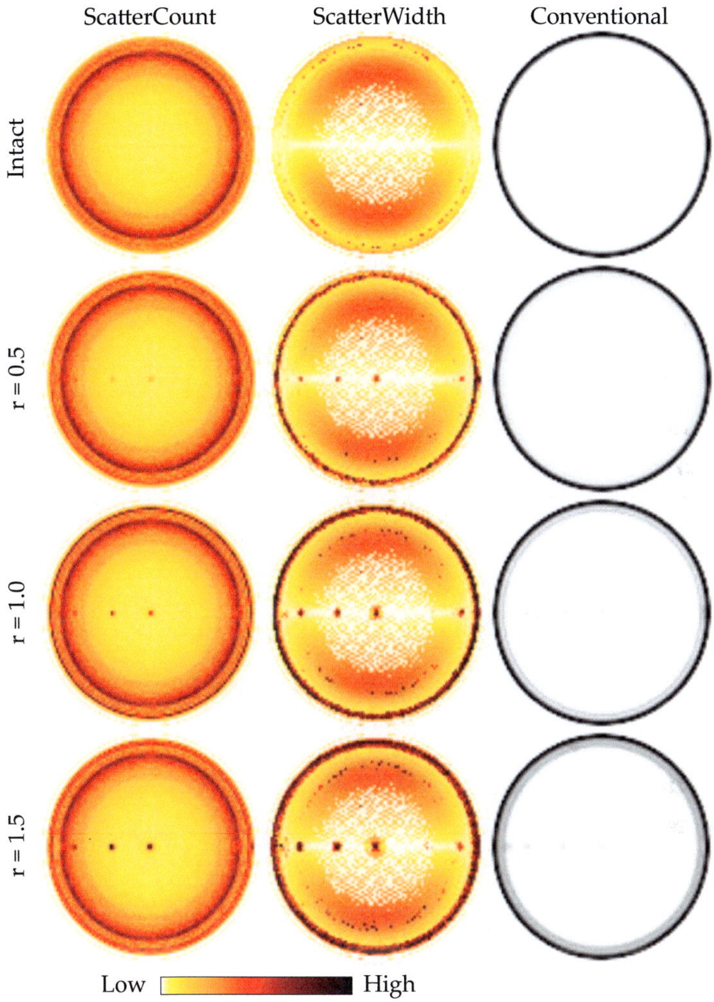

Figure 6.16: Results of the simulated experiments. The images corresponding to the two introduced features are shown in pseudo colors. The first row corresponds to the defect-free test object and the other rows to the test objects affected with scattering air bubbles. The parameter r denotes the defects' sizes as multiples of the projected area of a camera pixel.

6.4.3 Discussion

This section has introduced the concept of processing light transport matrices for the visual inspection of transparent objects. Two features have been described which can be extracted out of a light transport matrix and which indicate scattering structures, e.g., enclosed air bubbles, located inside a test object. Based on the performed experiments, the presented method is evaluated according to the criteria specified in Section 1.1.2:

1. **Versatility:** The experiments showed, that the introduced features allow to reveal scattering material defects inside transparent test objects having a complex geometry. Besides, the approach can make opaque defects visible by acquiring a bright field image for all pixels of the programmable light source turned on simultaneously.

2. **Accuracy:** The feature ScatterCount achieved CNRs for all defect sizes that were higher than those of the conventional inspection approach. Hence, a reliable detection of defects is possible.

3. **Speed:** For the parameters used in the performed experiments, the acquisition of the light transport matrix would require approximately 140 min and the processing time would be in the range of 18 s.

4. **Adaptability:** No adaptation of the system is needed for new types of test objects as long as their whole volume can be imaged by the telecentric camera system.

5. **Robustness:** The described method is robust against variations of the test object's geometry or its alignment in the measurement field since the introduced features are invariant with respect to such deviations.

Light transport matrix processing for the visual inspection of transparent objects is still in an early phase of development. Especially the long acquisition time renders this approach infeasible for practical applications. By employing the concept of compressed sensing, the sparsity of the light transport matrix could be exploited to greatly reduce the acquisition time.

6.5 Summary

This section presented three sets of methods for processing light fields acquired from transparent objects having a complex geometry in order to reveal light absorbing and light scattering material defects: methods for the processing of deflection maps, image fusion algorithms and an approach for processing light transport matrices.

For deflection maps, which can be acquired with a $4f$-light field camera, a schlieren deflectometer or a laser deflection scanner, a processing framework based on an adequate formulation of a spatial gradient has been introduced. By means of suitable distance functions, the values of the components of this gradient are able to indicate spatial discontinuities of deflection distributions of adjacent object points. Such discontinuities are a strong indication for scattering material defects. The Earth Mover's Distance and the Generalized Cramér-von Mises Distance have been adapted to serve as adequate distance functions. The fastCMD algorithm has been introduced which reduces the computational complexity of the Generalized Cramér-von Mises Distance from $\mathcal{O}(n^{2.5})$ to $\mathcal{O}(n^{1.5})$. Several experiments have been described and evaluated which state that the presented methods of acquiring and processing deflection maps are superior to state-of-the-art inspection systems with respect to the major relevant criteria.

A laser deflection scanner equipped with a PSD can measure the mean deflection direction at every measurement point. A method for processing these PSD-signals based on the calculation of its Jacobian and the Frobenius norm has been presented. By this means, larger scattering defects and some absorbing contaminants can be revealed in just 1.26 s per test object. Since this is noticeably faster than deflection map processing, it presents an alternative for inspection tasks requiring very short inspection times.

Two image fusion algorithms have been introduced that combine a series of images of a test object acquired under different illumination modalities to yield an inspection image in which material defects are highlighted and disturbing reflections are suppressed. The algorithms optimize an adequately formulated signal-to-noise ratio by exploiting information about the inspection system and the test object. The required information is acquired in an automated learning procedure by capturing images of a series of defect-free test objects and, if applicable, of a test object with artificially introduced defects. Several performed experiments showed that the approaches made defects visible which could not be revealed by a conventional inspection system.

The light transport matrix contains information about the optical paths of light rays propagating in a scene. For a transparent object that is placed between a telecentric camera and a programmable area light source, this information can be exploited to reveal scattering defects. For this purpose, two features have been defined that can be extracted out of the light transport matrix. These features yield high values for scattering material defects since they are sensitive to light rays that are scattered into multiple directions on their way from the light source to the camera. Although the approach is still in an early stage, simulated experiments stated its suitability for revealing scattering defects and that it can be superior to conventional methods.

7 Inverse light field illumination

The main physical effect affecting the light transport at and inside transparent materials is refraction. The transition between materials having different indices of refraction leads to changes of the direction of propagation of the light rays passing the boundary in between. As described in Chapter 1, scattering material defects also have an effect on the direction of propagation of the light. The effect of refraction can be inverted by emitting a specifically adapted light field. This technique, called inverse light field illumination, can be employed to compose a system for the visual inspection of transparent objects. In the following section, Section 7.1, the optical setup of the inverse light field illumination method is introduced. Section 7.2 presents the performed simulated and real experiments and their results. The topic is closed by a discussion of the suitability of the inverse light field illumination method as a visual inspection system given in Section 7.3 and a summary is provided in Section 7.4.

7.1 Optical setup

A visual inspection system based on inverse light field illumination relies on the concept of inverse light transport. The method is divided into two stages. In the first step, a reference light field is acquired from a defect-free instance of the object to be inspected. Figure 7.1 illustrates the concept of the respective optical setup. A transparent test object that is illuminated by collimated light transforms the incident parallel light beams into a light field L'. This light field is captured by a light field sensor positioned at a distance Δ behind the test object. In order to inspect further objects of the same type, the light transport is inverted as follows. A light field generator (cf. Section 3) is placed at the original position of the light field sensor at a distance of Δ from the test object. The light field generator emits $\widetilde{L}'$, which is the inverse of L' in the sense that the directions of propagation of the contained rays have been reversed. The other side of the object is observed by a telecentric camera system. In a telecentric lens, a so-called telecentric stop, i.e., an aperture with a small opening, is placed at the focal plane of the lens. Hence, the sensor captures only those light rays that propagate parallel to the optical axis and therefore pass the telecentric stop. Usually, telecentric systems are employed because the magnification of the optical system does not depend on the distance of the observed object, allowing accurate measurements. However, light field applications also benefit from telecentric systems, since the direction of propagation of the observed light bundles can be assumed to be parallel to the optical axis. Regions of the test object which are completely free from any material defects transform the incident light field $\widetilde{L}'$, emitted by the light field generator, to $\widetilde{L}$, i.e., light rays propagating parallel to the optical axis in the direction of the telecentric camera (see Fig. 7.2, blue optical path). These parallel rays pass the telecentric stop and therefore produce a high signal level captured by the sensor, which thus indicates a defect-free region of the test object. If a material defect is present inside the test object, that leads to light rays being absorbed (e.g., by opaque contaminants) or deflected (e.g., by scattering inhomogeneities), the ray will not contribute to a signal on the sensor. In case of an absorbing defect, the light is directly eliminated and does not reach the camera. Scattering defects will deflect the affected light rays, so that they no longer propagate parallel to the optical axis, and thus get blocked by the telecentric stop (see Fig. 7.2 red optical path). For

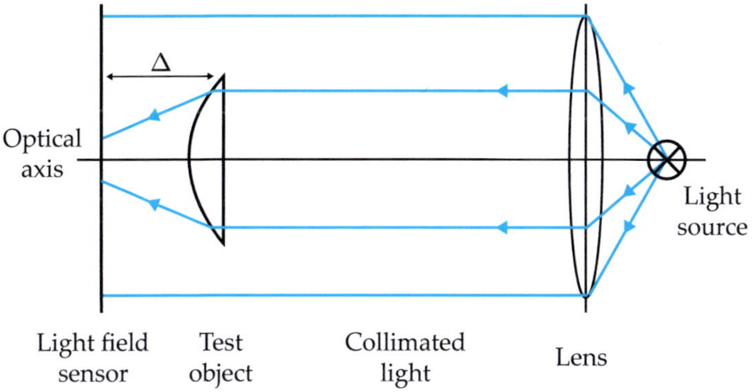

Figure 7.1: Acquisition of the reference light field L'. The test object transforms a collimated light beam into the light field L', which is captured by a light field sensor.

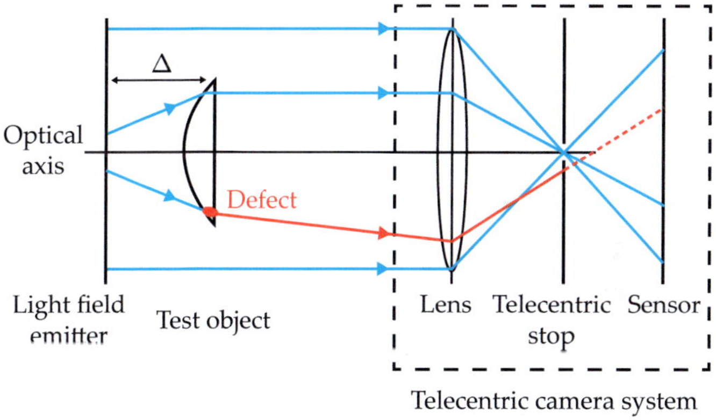

Figure 7.2: Inspection by inverse light field illumination. A light field emitter placed at the original position of the light field sensor emits $\widetilde{L}'$. Defect-free regions of the test object transform the incident rays on the left side into rays running parallel to the optical axis on the right side. Parallel rays pass the telecentric stop of the telecentric camera system and yield a high signal on the sensor (blue optical path). All other rays are blocked by the telecentric stop and result in a low signal value, indicating a material defect (red optical path).

many object geometries, this approach allows inspecting the whole volume of the test object by acquiring only a single image and by searching for pixels having low gray values.

The next section, Section 7.2, will cover both simulated and real experiments evaluating the presented inspection method.

7.2 Experiments

In order to evaluate the inverse light field illumination method, simulated experiments have been performed by means of the simulation framework introduced in Chapter 5 and real experiments based on the light field generator prototype presented in Chapter 3 have been conducted. The goal of these experiments is to show the principle suitability of the presented approach for the visual inspection of transparent objects and to provide a proof of its practical feasibility. Hence, the following discussions focus on a qualitative evaluation.

7.2.1 Simulated experiments

The introduced simulation framework (see Chapter 5) allows evaluating the presented approach by means of simulated experiments. Therefore, different virtual scenes have been designed and rendered. A double-convex lens has been employed as the test object. Adequate modifications applied to the test object realize different types of material defects. Besides the inverse light field illumination setup, two conventional inspection approaches employing a collimated light source and an area light source have also been simulated for every instance of the test object. The same telecentric camera was used in all three configurations. By means of the resulting simulated images, the different inspection methods will be compared with each other.

7.2.1.1 Virtual scenes

The following paragraphs describe the simulated virtual scenes and discuss the results. Each simulation treats a different kind of material defect affecting the test object.

Defect-free test object The simulation results for a defect-free test object are shown in Fig. 7.3. The parallel rays emitted by the telecentric illumination are deflected by the test object and therefore do not propagate parallel to the optical axis after passing the test object. Light bundles captured by the telecentric camera have only small deflection angles with respect to the optical axis since otherwise they would have been blocked by the telecentric stop. The degree to which rays can be deflected while still being captured depends on the diameter of the telecentric stop. In the simulated case, only rays passing the test object near its center propagate with deflection angles which are small enough to be captured by the camera.

The simulated image resulting for the area illumination shows a constantly high signal. The area illumination emits light rays in multiple directions. Although these rays get deflected by the test object, for each object point there are still some rays propagating parallel to the optical axis, so that light shines on every pixel of the telecentric camera system.

The results for the inverse light field illumination are as expected. The test object transforms the emitted light field into rays propagating parallel to the optical axis, leading to high signal values for nearly all pixels. There are some artifacts visible at the borders of the test object. This can be explained by considering the respective rays of sight of the telecentric camera. At these object positions, the rays of sight get strongly deflected so that they hit the light field generator at angles of incidence for which it emits no light.

Telecentric illumination

Area illumination

Inverse light field illumination

Figure 7.3: Simulated images of a defect-free test object.

Absorbing contaminants Two absorbing, i.e., opaque spheres are placed at different positions inside the test object to simulate absorbing inclusions. One defect is located at the center of the test object and the other defect is placed at the same depth but moved towards the bottom border of the object. Figure 7.4 shows the resulting simulated images. The collimated illumination reveals only the center defect. This is because in this region the deflection of the parallel rays is still small enough that they are captured by the telecentric camera. The second defect cannot be seen due to the refraction of the test object.

The area light source is able to make both absorbing contaminants visible. The two defects block the optical paths from the rays of sight of the camera to the area light source, resulting in the sensor observing low intensity values. Since the area light source emits light diffusely in many directions, the whole test object can be tested for absorbing defects.

Both defects are also made visible by the inverse light field illumination. At the positions of the contaminants, the light rays of the incident light field are absorbed and can therefore not be transformed into parallel rays that would be captured by the telecentric camera. This results in dark pixels in the camera image at the positions corresponding to the two inclusions.

Scattering contaminants By positioning two small spheres with an index of refraction of 1.0 inside the test object, scattering material defects (e.g., enclosed air bubbles) are simulated. The defects are located at the same positions as were the absorbing defects, i.e., at the center of the object and near its border at the bottom. Figure 7.5 shows the resulting simulated images. Both defects are made visible by the collimated illumination. The defects scatter the incident light rays in various directions, so that some rays propagate parallel to

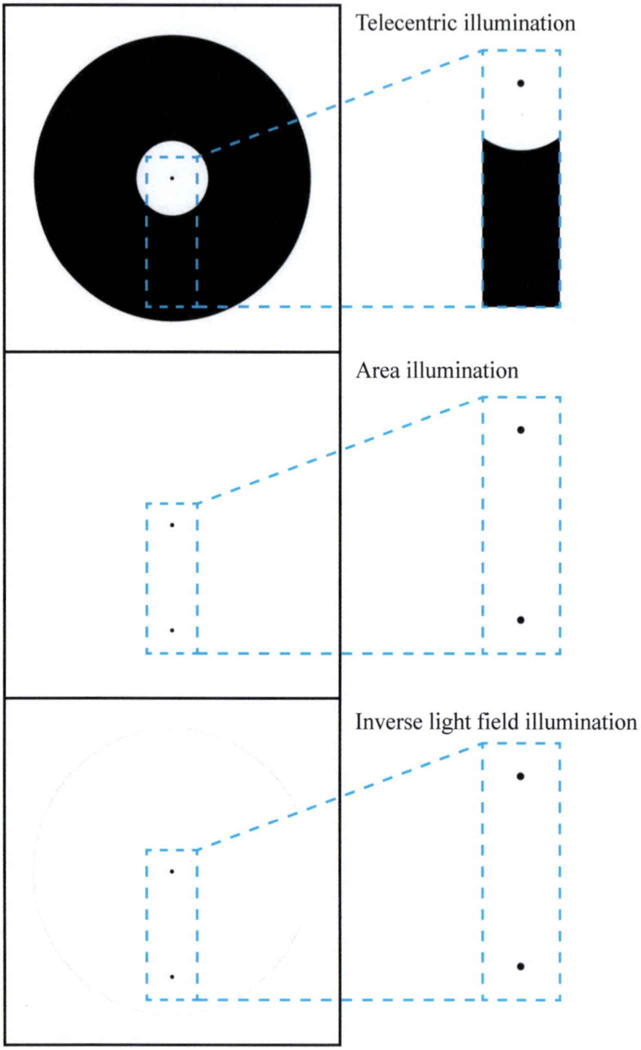

Figure 7.4: Simulated images of a test object affected by two absorbing impurities. The highlighted regions are magnified on the right-hand side.

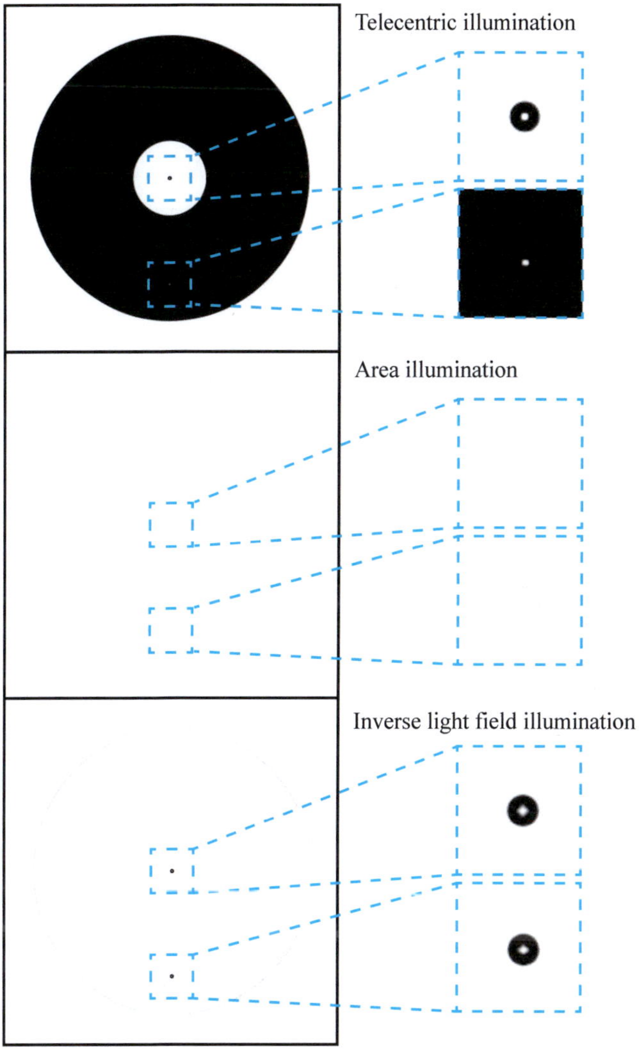

Figure 7.5: Simulated images of a test object affected by two scattering impurities. The highlighted regions are magnified on the right-hand side.

the optical axis and are therefore captured by the telecentric camera. However, this illumination only allows correctly inferring the size of the center defect (dark region including white spot). In contrast, the image of the defect near the border of the object consists of only some white pixels inside a region that is smaller than the actual defect.

In the images resulting from the area illumination, the defects are barely visible. Only the interfaces of the defects with the surrounding test object appear, but with low contrast. This is because the scattering caused by the defects does not greatly affect the distribution of the directions of propagation of the illuminating light rays. The interfaces of the defects result in strong deflections, directing some light out of the system, thus yielding the faint images of the defects.

The test object transforms the inverse light field emitted by the light field generator into parallel rays everywhere except at the positions of the two air bubbles. The defects scatter the incident rays in multiple directions, so that they do not propagate parallel to the optical axis, and hence are blocked by the telecentric stop of the camera. For this reason, the defects are made clearly visible. Furthermore, it is also possible to estimate the sizes of the defects based on the image obtained using the inverse light field illumination.

Incomplete shape An incomplete shape, as could be caused, e.g., by an imperfect plastic molding process, has been realized with the principle of constructive solid geometry (CSG). The CSG method is a common tool in the field of computer graphics, and is used to create new shapes by combining simple shapes with Boolean operators. With CSG, parts of two spheres with different sizes are removed from the original test objects by means of the Boolean difference operator. Figure 7.6 illustrates the two modeled test objects.

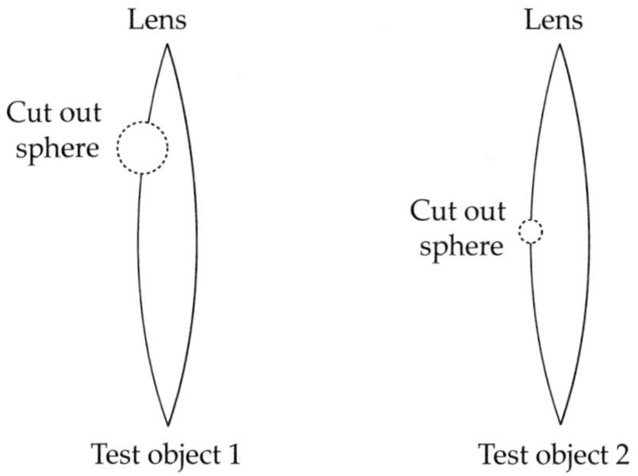

Figure 7.6: Modeling of incomplete shapes. By employing constructive solid geometry, two spheres of different sizes are cut out of the original test object.

The inspection images resulting from the first incomplete test object are shown in Fig. 7.7. The collimated illumination reveals the defect. However, as with the scattering defects, it is not possible to judge the correct size of the defect. This is because only some parts of the defective region transform incident light into rays that propagate parallel to the optical axis and get captured by the telecentric camera.

The inspection image for the area light source also shows the missing part of the object. Yet only the borders of the defective region lead to strong deflections directing incident light rays out of the optical system. Hence, only the borders of the defect and not its complete region is made visible. Indeed, the image allows measuring the size of the defect.

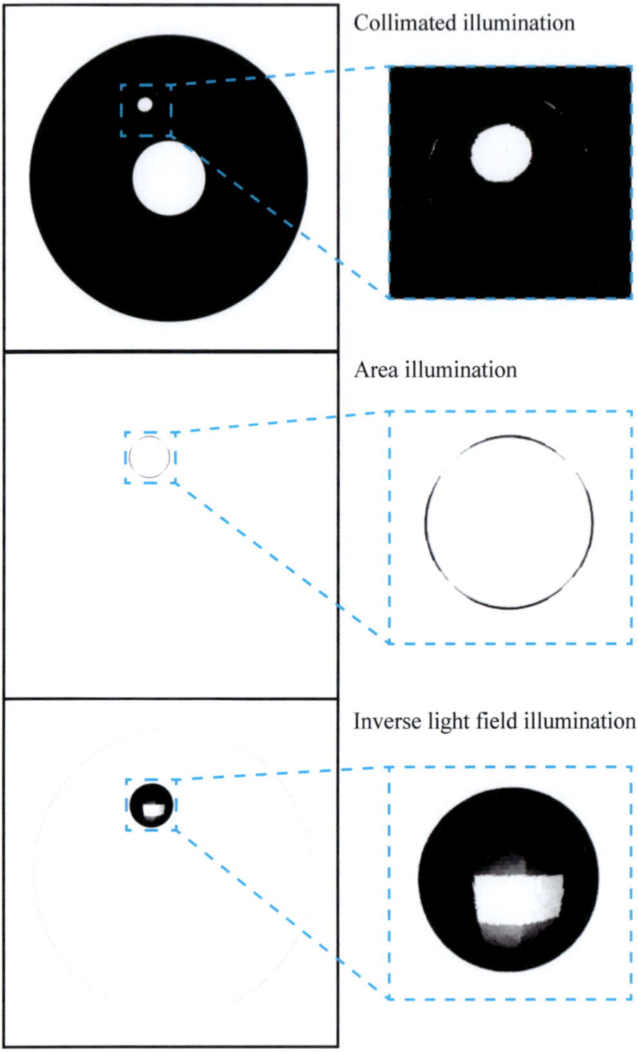

Figure 7.7: Simulated images of a test object with a strongly incomplete shape (corresponding to test object 1 in Fig. 7.6). The highlighted regions are magnified on the right-hand side.

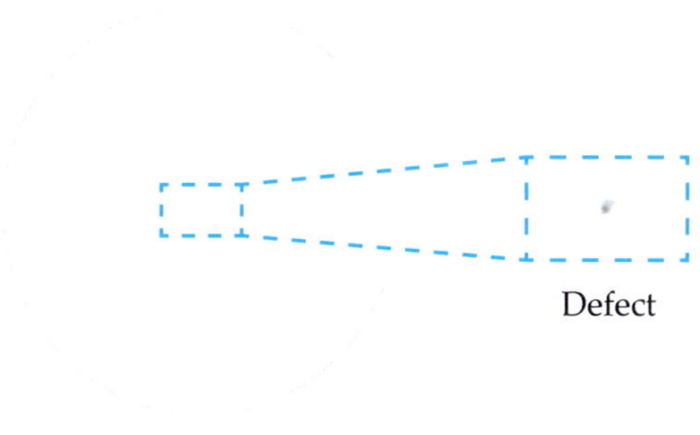

Figure 7.8: Simulated inspection image of the inverse light field approach for a test object whose shape is missing a section of small sphere (corresponding to test object 2 in Fig. 7.6).

The defect is made clearly visible by the inverse light field illumination. The affected region of the test object has different surface normals than the defect-free object, so that the incident light rays are not transformed into rays propagating parallel to the optical axis. These rays are blocked by the telecentric stop, which results in pixels with low gray values. However, since the surface profile of the defect is shaped like a part of a sphere, it has a magnifying optical effect. This explains the grid-like structure that can be seen in the defective region, which shows parts of the pixels of the light field generator. The inverse light field illumination also allows inferring the size of the defect.

Figure 7.8 shows the result for the second test object (cf. Fig. 7.6). Only the result for the inverse light field illumination is provided, since the other two approaches did not make the defect visible at all. The inverse light field illumination clearly reveals the small shape defect.

Deviations of the index of refraction High precision optical systems require the employed optical elements to have a homogeneous distribution of the index of refraction. Differences in the index of refraction from its specified value can also be considered as material defects. This has been simulated by increasing the index of refraction of the test object by 10% compared to the original test object. Figure 7.9 shows the resulting simulated images.

For the collimated illumination, the radius of the bright circle in the center is smaller compared to the defect-free test object (cf. Fig. 7.3). Due to the higher index of refraction, the deflection of light rays passing the test object is increased, so that those rays which are transmitted by the test object further away from its center are blocked by the telecentric stop of the camera system. Hence, a detection of this defect is not reliably possible by means of a collimated illumination.

The increased deflection angles caused by the higher index of refraction result in more of the light rays of the area illumination completely missing the camera. This is why the inspection image acquired using the area illumination shows the test object with slightly reduced gray values. This indicates that the test object is affected by some sort of defect.

The inspection image obtained with the inverse light field illumination clearly shows that the whole volume of the test object differs from its specifications. Only bundles of rays passing the center of the object are still completely captured by the telecentric camera since they are not deflected while passing through the test object. At all other positions, the increased index of refraction results in different deflection angles compared to the original test object, so that the transmitted light rays do not propagate parallel to the optical axis, which means they get blocked by the telecentric stop. The grid-like pattern of the inspection image most likely shows a magnified part of the pixel structure of the light field generator.

Figure 7.9: Simulated images of a test object having an index of refraction that is 10% higher than specified.

Misaligned position The correct positioning of components is crucial for many optical systems. Since all three considered concepts are able to make visible misalignments perpendicular to the optical axis, the test object is moved along the optical axis towards the camera in the simulated virtual scene. The displacement does not exceed the depth of field of the optical system, so that focused images are acquired. Figure 7.10 shows the simulation results.

The inspection image obtained using the collimated illumination does not differ from the image corresponding to a defect-free object. The changed position of the object does not affect the parallel rays of sight of the camera system. Hence, an axial shift of the test object within the depth of field cannot be detected by this illumination.

Neither does the area illumination reveal this simulated misalignment. Indeed, at every object position there is an incident illuminating ray which propagates parallel to the optical axis after passing through the test object. Hence, the complete test object appears the same as the defect-free instance.

In contrast to the two other types of light sources, the inspection image of the inverse light field illumination clearly differs from that of a correctly positioned test object. The emitted reference light field does not fit the (mis-) alignment of the object, leading to nearly all the transmitted rays propagating in directions that are not parallel to the optical axis. Therefore, the whole image of the test object indicates a difference of the test object from the intended specifications. As for the case of the changed index of refraction, the pattern that is visible in the inspection image is a magnification of the pixel structure of the light field generator.

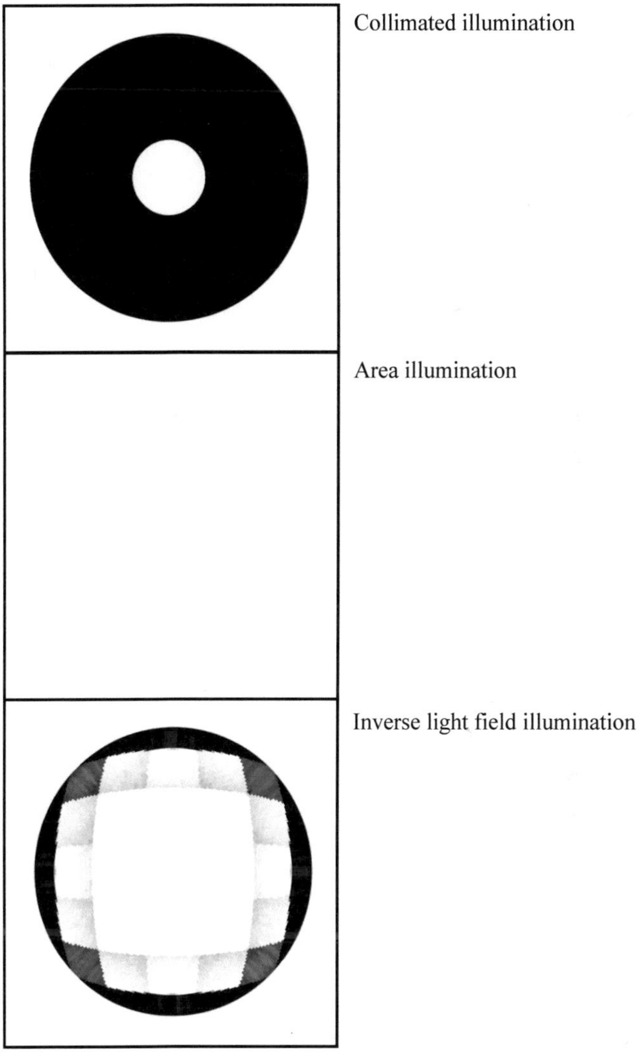

Figure 7.10: Simulation images of a test object that is moved along the optical axis away from its intended position.

7.2.2 Real experiments

Additional real experiments have been conducted to evaluate the practical feasibility of the method. A double-convex lens and small glass bottles served as test objects. The prototype of the light field generator described in Section 3.3 was employed to emit the inverse light field. The image acquisition has been performed with a telecentric camera system.[1]

7.2.2.1 Acquisition of the inverse light field

In contrast to the simulated experiments, the inverse light field $\widetilde{L}'$ was not obtained via capturing the reference light field L' with a light field sensor. The optical setup used for the inspection phase is directly suitable for acquiring $\widetilde{L}'$ by means of an illumination series. The light field generator is based on a two-dimensional high resolution display in concert with a microlens array as introduced in Chapter 3. Hence, every one of the pixels of the display allows the emission of a bundle of approximately parallel light rays originating from the corresponding spatial location on the light field generator and propagating in the corresponding direction. When a defect-free test object is placed between the light field generator and the telecentric camera, the inverse light field can be obtained using Algorithm 3.1 which has been presented for capturing the set of central pixels in the calibration of the light generator (see Section 3.2). This algorithm determines all pixels of the light field generator that lead to light bundles which are incident to the telecentric camera with a direction of propagation that is parallel to the optical axis. Exactly these pixels correspond to the sought inverse light field $\widetilde{L}'$.

[1] More details about the used components are provided in Appendix A.3.

7.2.2.2 Test object 1: Double-convex lens

A double-convex lens has been employed as a first test object for the empirical experiments. The inverse light field $\widetilde{L}'$ is obtained using Algorithm 3.1. Inspection images have been captured for the lens residing in the position of the acquisition of $\widetilde{L}'$, for the lens being completely removed, for the lens being rotated by 2° around the horizontal axis perpendicular to the optical axis, and for the lens being shifted by 2 mm to the left along the horizontal axis. Figure 7.11 shows the corresponding camera images. For the defect-free case, i.e., the lens being in the position of the acquisition of the inverse light field $\widetilde{L}'$, the majority of the pixels inside the image region covered by the test object show high gray values, as expected. However, there are also many pixels that have lower gray values or are even completely black. This is due to the shortcomings of the light field generator employed. For the dark camera pixels, there are no corresponding combinations of pixels and microlenses of the light field generator that can emit a bundle of rays in the correct direction. Accordingly, the respective parts of the test object cannot be inspected.

The inspection images nevertheless show the potential of the proposed approach. If the lens is missing, nearly the complete inspection image shows gray values near zero. There are some bright pixels in the center since at these positions, the inverse light field contains rays parallel to the optical axis. This is because the surface of the test object is aligned perpendicular to the optical axis in its center. As these rays propagate parallel to the optical axis also when no test object is present, they result in high signals at the respective pixel positions.

Rotating the lens by 2° also results in an inspection image with nearly no high signal values, indicating that the test object differs from its specifications. Due to the rotation, no light rays of the inverse light field propagate parallel to the optical axis after passing through the test object, and are therefore blocked by the telecentric stop of the camera.

A horizontal shift of the lens by 2 mm to the left along the horizontal axis perpendicular to the optical axis leads to a similar inspection image as that from the rotation. The resulting image has nearly no pixels with high gray values. Because of the shift of the lens, the transmitted light rays do not propagate parallel to the optical axis. This results in low signal values at the corresponding pixels. Again, this clearly indicates that the test object is misaligned or that its complete volume is affected by an anomaly.

7.2.2.3 Test object 2: Glass bottles

For the second experimental setup, two small glass bottles were used, which are usually employed as containers for liquid drugs. Again, Algorithm 3.1 was utilized to acquire the inverse light field $\widetilde{L}'$. Inspection images were captured for the reference bottle in its original position, for the reference bottle shifted horizontally by 2 mm, for the reference bottle being missing and for the second bottle being placed at the position of the reference bottle. The glass of the second bottle is slightly thicker, so that it would be a wrong bottle type in the context of a production environment. Figure 7.12 shows the resulting inspection images.

For the reference bottle residing in its intended position, the image region of the body of the bottle shows mostly bright pixels. At these positions, the rays of the inverse light field are transformed into rays propagating parallel to the optical axis, indicating that the object meets the specifications. However, the pixels corresponding to the neck of the bottle show almost only low gray values. This is because the curvature of the glass surface is so strong at these positions that the light field generator is not able to emit rays with suitable directions of propagation. Consequently, the corresponding regions of the test object cannot be completely inspected.

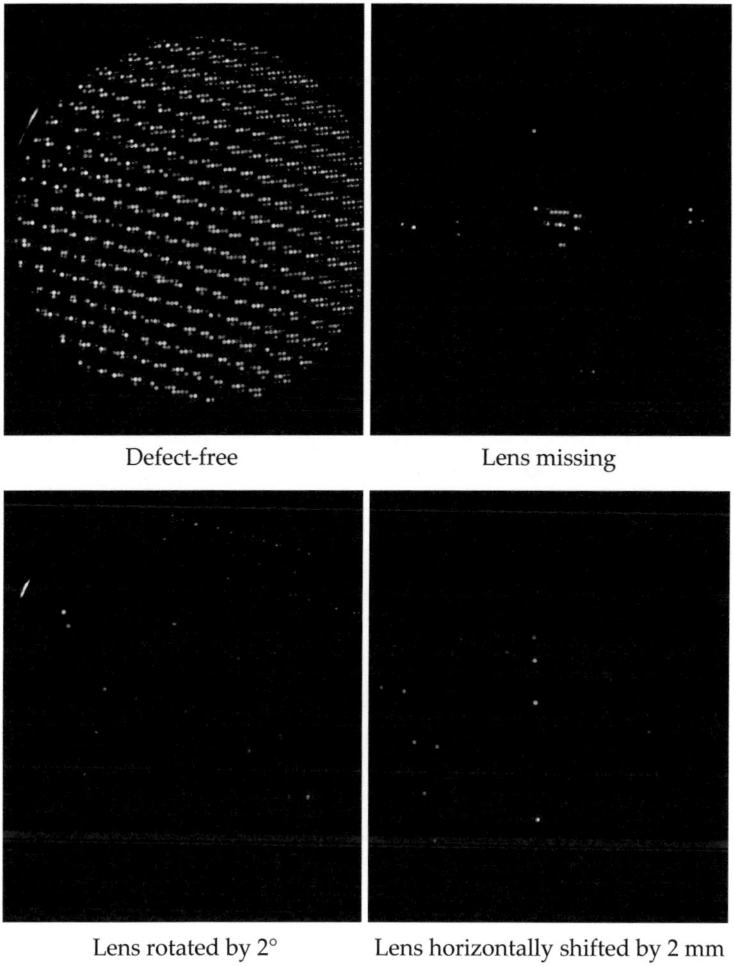

Figure 7.11: Inspection images resulting for the experiments performed with the double-convex lens.

When the reference bottle is missing, the corresponding image regions show gray values near zero, except for the center part of the body of the bottle. At these positions, the inverse light field already contains rays propagating parallel to the optical axis, since there the surface of the bottle would be oriented perpendicular to the optical axis. Hence, these rays result in some bright pixels in the corresponding region even if the bottle is absent.

The inspection image for the second, slightly thicker bottle type mostly consists of nearly black pixels in the image region of the test object. The increased thickness of the glass and the slightly different shape result in directions of propagation of the transmitted light rays that are not parallel to the optical axis. The rays are blocked by the telecentric stop, revealing that the bottle is not as intended. This approach should also result in low-value pixels if the glass bottle were to be filled with the wrong liquid (if the reference liquid and the wrong liquid had different indices of refraction).

A horizontal shift of the reference bottle by 2 mm also results in an inspection image that is nearly completely black in the region of the test object. This is again caused by the rays being deflected into directions not parallel to the optical axis.

7.3 Discussion

The results of the performed experiments can be summarized by means of the criteria described in Section 1.1.2 regarding a visual inspection system:

1. **Versatility:** Opaque and scattering material defects hidden inside a transparent object with a complex shape can be made visible.

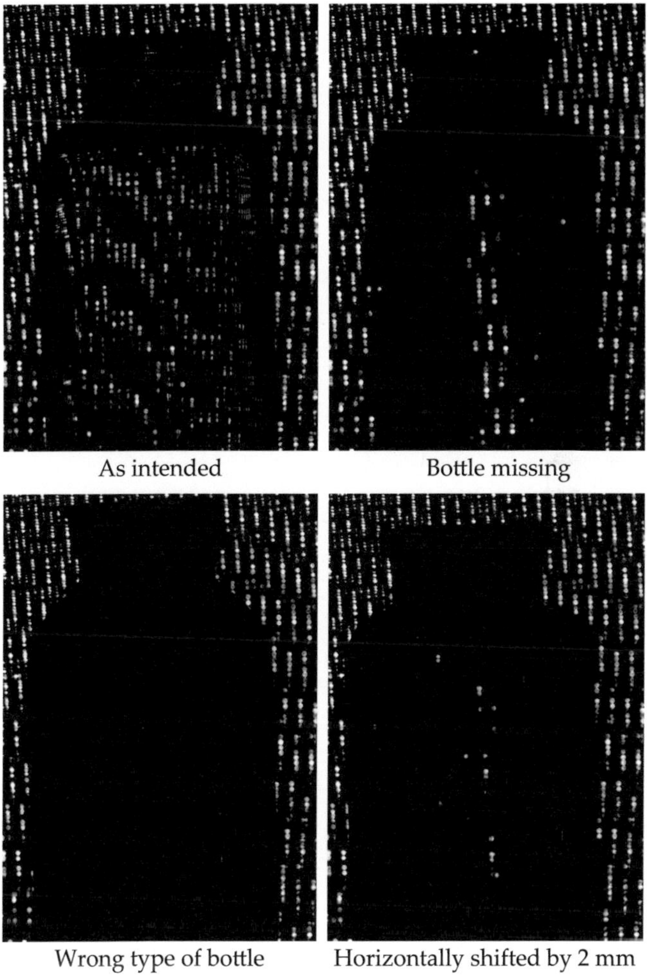

Figure 7.12: Inspection images resulting for the experiments performed with the glass bottles.

2. **Accuracy:** Based on the current state of the method, the accuracy can not be quantitatively evaluated. However, the experiments show that defects are made visible with high contrast and that nominal object structures are suppressed.

3. **Speed:** As soon as the inverse light field corresponding to a test object type has been acquired, the complete inspection of a test object takes as long as capturing a single image. Depending on the employed camera system, the required inspection time will be in the range of 1 ms.

4. **Adaptability:** In order to adapt the system to a new type of test object, the corresponding inverse light field has to be acquired. Assuming a camera with a resolution of $5,000 \times 4,000$ = 20 mega pixels, $\log_2(5,000) + \log_2(4,000) \approx 25$ images would have to be captured when using the acquisition approach based on the coded illumination series. For a state-of-the-art camera with a acquisition rate of 30 Hz this would require less than one second.

5. **Robustness:** As shown by the real experiments, the approach is very sensitive to small variations of the position and the orientation of the test object.

7.4 Summary

This chapter introduced the method of inverse light field illumination, which allows visually inspecting a transparent test object by capturing and evaluating a single image only. By means of Algorithm 3.1, the system can be adapted to a new type of test object without the user having to create a new specific system design. During the inspection process, test objects are illuminated with the acquired reference light field by a light field generator. Since the light field is inverse to the intended one, i.e., the nominal object state, a defect-free object will transform the incident light field into collimated light propagating parallel to the optical axis whereas material defects will disturb the transmitted light field. The light field exiting the investigated test object is observed by a telecentric camera system so that ray bundles not propagating parallel to the optical axis, i.e., that correspond to material defects, will result in image regions of low intensities. Consequently, a computationally simple thresholding operation of the acquired image is sufficient to detect material defects. Similar to most inverse illumination techniques, this method is also very sensitive to the geometric arrangement of the inspection system and the test object. This can represent an advantage if verifying the arrangement is part of the inspection task, but could instead be problematic since positioning errors can lead to falsely rejected test objects.

The inspection method was evaluated by simulated experiments using the Mitsuba renderer and by real experiments based on the developed light field generator prototype. The simulated experiments clearly show that the inverse light field illumination method can outperform conventional inspection approaches. The real experiments confirm the practical feasibility of the method but also show that its accuracy greatly depends on the employed light field generator. Hence, for the current state of technology, inverse light field illumination is not applicable to high-precision inspection tasks. Nevertheless, since light field generators represent an active research topic, better realizations can be expected in the near future.

8 Summary and outlook

In this thesis, the problem of the automated visual inspection of transparent objects having a complex geometry has been studied with the focus laid on light field methods. Since the test object itself and contained material defects influence the intensity and the direction of propagation of transmitted light rays, considering both the light's spatial and directional properties is crucial for finding material defects. Designing a conventional inspection system for complex-shaped transparent objects requires a tremendous amount of expertise and design time in order to succeed in detecting absorbing and scattering contaminants. This effort of adapting an inspection system to a new type of test object is greatly reduced by the contributions of this thesis. For all major parts of an automated visual inspection system, the illumination component, the sensor component and the signal processing unit, novel methods and approaches for emitting, capturing and processing light fields have been presented. The contributions constitute a toolbox to develop new visual inspection systems for uncooperative transparent objects. By means of simulated and real experiments based on prototypes, the methods have been evaluated and their superiority over conventional inspection systems could be proved with respect to several criteria.

On the next two pages, Tab. 8.1 provides an overview over the introduced inspection methods and their evaluation with respect to the state of the art. The single contributions are summarized in Section 8.1 and Section 8.2 presents topics and possible ideas concerning future research.

	Deflection map processing	SNR-based image fusion
Versatility	Opaque defects Scattering defects	Opaque defects Scattering defects
Accuracy	High	Very high Suppression of disturbing reflections
Speed	Acquisition time: Schlieren deflectometer: ~ 3.0 s Laser deflection scanner: ~ 5.5 s Processing time: ~ 2.2 s (Depending on resolution)	Acquisition: ~ 2 s Processing: ~ 1 s
Adaptability	No adaptation required	Set of defect-free objects Signal sample with artificially introduced defects
Robustness	Variations of test object's geometry or misalignment irrelevant	Robust to small variations of test object's geometry Sensitive to misalignments of the test object

Table 8.1: Overview of the evaluation of all presented light field methods for the visual inspection of transparent objects. The colors of the cells indicate the degree to which a criterion has been met compared to the other methods (green: best, yellow: medium, red: worst).

Inverse light field illumination	Conventional bright field	Conventional bright field + dark field
Opaque defects Scattering defects	Opaque defects only	Opaque defects Scattering defects
Very high	Low	High Sensitive to disturbing reflections
Acquisition: ~ 0.075 s	Acquisition: ~ 0.075 s	Acquisition: ~ 2 s
One defect-free object required	No adaptation required	Expert needed for revision of system design
Sensitive to variations of test object's geometry and its misalignment	Variations of test object's geometry or misalignment irrelevant	Small variations of test object's geometry tolerable Sensitive to misalignments of the test object

(Table 8.1 continued)

8.1 Conclusive evaluation of contributions

The laser deflection scanner in concert with the algorithms for the processing of light deflection maps, the image fusion methods optimizing the signal-to-noise ratio to obtain improved inspection images and the inverse light field illumination approach by means of a light field generator constitute the main contributions of this thesis. In the following, the individual contributions are summarized and their suitability for a visual inspection system in the sense of the criteria introduced in Section 1.1.2 is recapitulated. Table 8.1 provides an overview of the evaluation. The introduced methods are compared with a conventional bright field setup, as it has been used in the experiments for the processing of deflection maps, and with an inspection system employing bright field and dark field illumination (e.g., the Purity system).

The novel sensor concept of the laser deflection scanner enables to acquire the angular deflection distribution of collimated light rays after being transmitted through a transparent test object. The sensor is not affected by the unfortunate tradeoff between spatial and angular resolution of conventional light field cameras. Both resolutions can be scaled accordingly with respect to the application. An additional position sensitive detector allows to capture the mean deflection direction for every measurement point. Another benefit of the laser deflection scanner is its capability of being installed as an in-line inspection system what eases its introduction in an existing production line.

A gradient has been formulated for detecting discontinuities in deflection maps between the deflection distributions of spatially adjacent measurement points. For this purpose, the Generalized Cramér-von Mises Distance has been adapted and employed as a suitable distance function. A fast algorithm, fastCMD has been developed, which reduces the computational complexity of the Generalized Cramér-von Mises Distance from $\mathcal{O}(n^{2.5})$ to $\mathcal{O}(n^{1.5})$ leading to feasible inspection times in the context of industrial inspection. Employing the laser deflection scanner in concert with the fastCMD-algorithm yields an inspection system capable of imaging opaque and scattering defects in complex-shaped transparent objects with high accuracy and with an acceptable inspection speed. Furthermore, the system can effortlessly be adapted to new types of test objects and it is robust against small shape variations and misalignments of the investigated object.

A method has been presented, with which the mean deflection directions captured by the position sensitive detector of a laser deflection scanner can be processed to reveal larger scattering defects and some kinds of absorbing contaminants. Since transferring the corresponding signals and calculating the involved Jacobian and Frobenius norm requires only 1.26 s per test object, this represents an adequate alternative to deflection map processing for inspection tasks requiring short inspection times.

Two image fusion algorithms have been introduced which can be employed to enhance the inspection images acquired with a conventional multi-channel dark field inspection system. A signal-to-noise ratio is formulated in which disturbing reflections caused by the surface of the test object constitute the noise component and the capability of the employed light sources to reveal defects represents the signal component. The image fusion methods optimize this signal-to-noise ratio in order to yield inspection images in which material defects are revealed and disturbing reflections are suppressed. An automated learning procedure allows to acquire information about the noise and the signal component based on a set of defect-free test objects or, if available, by means of a test object with artificially introduced defects. The approach greatly improves the adaptability with respect to the conventional system at the minor cost of an additional processing time of one second.

The presented method of inverse light field illumination allows to inspect a transparent object by acquiring a single image only. The test object is illuminated with an inverse light field that inverts the refraction effects caused by the object. If the object is free from defects, the incident inverse light field is transformed into light rays propagating parallel to the optical axis. A telecentric camera, which captures only such rays, observes the other side of the test object. Material defects inside the test object will either absorb incident light rays or change their direction of propagation, so that they will be revealed as regions of low intensity in the captured image. Noticeably, no further processing is required to detect the defects. The idea of inverse light field illumination could be realized by means of the presented concept of a light field generator. A high resolution programmable two-dimensional display that is positioned in the focal plane of a microlens array allows to emit customized light fields. The constructed prototype of the light field generator allowed to prove the suitability of the inverse light field illumination method for the visual inspection of transparent objects. The achievable accuracy is currently impaired by the limitations of the light field generator caused by its early stage of development. However, advanced technologies for the emission of light fields can be expected in the near future. The sensitivity of the approach with respect to variations and misalignments of the test object represents a disadvantage of the method. Compared to the other approaches, it achieves the fastest inspection speed and it can easily be adapted to new types of test objects by acquiring the corresponding reference light field.

A conceptual approach for inspecting transparent objects based on processing their light transport matrix has been introduced as a further contribution. An optical setup for the acquisition of light transport matrices and the corresponding processing methods for revealing scattering material defects have been described. The light transport matrix contains the information about the contribution of every light source of a scene to every detector element of the employed camera. Two features have been presented which can be extracted out of the light transport matrix and which are sensitive to scattering defects. The features are invariant to the geometry of the test object and to its position and orientation in the measurement field. With this approach, test objects with a complex shape can be inspected, the method is robust against small variations and misalignments of the test object and no adaptation is needed for new types of test objects. However, in order to transfer the concept to an actual inspection system, the time for the acquisition of the light transport matrices has to be greatly reduced.

8.2 Future work

The presented contributions provide several opportunities for conducting future research. The internal sensor design of the laser deflection scanner could be updated, e.g., by employing suitable FPGAs (field programmable gate arrays), in order to realize shorter acquisition times. Furthermore, adequate processing of the acquired light deflection maps would allow to obtain spatially resolved estimations of the optical power of transparent test objects. By this means, the distribution of the index of refraction of a test object could be reconstructed if its surface geometry is known. Methods of algorithm engineering could be applied to further speed up the computation of the Generalized Cramér-von Mises Distance, e.g., by exploiting the massive parallelism of a graphics processing unit. This could reduce the inspection time and hence resolve the remaining drawback of the methods for processing deflection maps.

The optical realization of the light field generator could be improved by better matching the periods of the microlens array and of the pixels of the underlying display. Also the focal length of the microlenses could be adjusted so that larger emission angles are possible. Generally, it might be interesting to optimize the optical parameters of the light field generator with respect to an actual inspection task (e.g., to the test object). By this means, the accuracy of the inverse light field illumination method could be increased.

With respect to the image fusion approaches, more information, e.g., the variance of the noise signal, could be exploited. The additional information could be employed to automatically infer a spatially resolved threshold to realize a robust detection of defects.

Another interesting future research topic is to evaluate whether methods from the field of compressed sensing can be used to speed up the acquisition of light transport matrices. Since these matrices are sparse, i.e., most of their elements are approximately equal to zero, noticeably fewer images should be sufficient for their reconstruction.

With respect to the common structure of a machine vision system (as illustrated by Fig. 1.1), the methods introduced in this thesis yield preprocessed inspection images in which material defects are highlighted and all other influences are mitigated. Hence, sensible future research could deal with methods and algorithms for further processing and classifying these images, e.g., by means of machine learning approaches.

A Appendix

A.1 Asymptotic notation of computational complexities

The so-called Big O notation allows to characterize the asymptotic computational complexity of algorithms [Meh08]. The set

$$\mathcal{O}(f(n)) = \{g(n) : \exists\kappa > 0 : \exists n_0 \in \mathbb{N} : \forall n \geq n_0 : g(n) \leq \kappa \cdot f(n)\} \tag{A.1}$$

contains all functions $g(n) : \mathbb{N} \to \mathbb{R}$, whose asymptotic behavior is bounded by a function $f(n) : \mathbb{N} \to \mathbb{R}$. If the complexity of an algorithm is in $\mathcal{O}(f(n))$, its asymptotically required number of computation steps for a problem instance of size n will be less or equal to $\kappa \cdot f(n), \kappa > 0$.

A.2 Measurement of the runtime of fastCMD

To empirically evaluate the speedup achievable by employing the fastCMD-algorithm with respect to the naive calculation of the CMD, the following experiment has been performed.

Multiple pairs of square two-dimensional histograms with numbers of bins $n \in [10 \cdot 10, \ldots, 100 \cdot 100]$ have been randomly generated and used as inputs to the algorithms. Both algorithms have been implemented in C# using the Accord.NET framework [Sou] and by not making use of any parallelism. For each histogram size, the measurement has been repeated 10 times. The results are listed in Tab.A.1 and clearly support the theoretically determined reduction of complexity from $\mathcal{O}(n^{2.5})$ to $\mathcal{O}(n^{1.5})$.

Table A.1: Execution times resulting from the performed experiment. The size of the input histograms is denoted by n and t_{naiv}, t_{fast} represent the mean measured execution times in milliseconds $\pm$ standard deviation for the naive, respectively, the fast algorithm.

$n =$	10^2	20^2	30^2	40^2	50^2
t_{naiv} in ms	7.2	183.5	1,311	5,379	16,145
	± 0.6	± 0.3	± 2	± 4	± 36
t_{fast} in ms	0.23	0.75	2.18	4.96	9.7
	± 0.02	± 0.01	± 0.01	± 0.02	± 0.3

$n =$	60^2	70^2	80^2	90^2	100^2
t_{naiv} in ms	49,379	105,820	201,940	375,010	634,180
	± 138	± 218	± 1,550	± 1,433	± 390
t_{fast} in ms	16.29	25.7	38.35	54.6	75.1
	± 0.06	± 0.1	± 0.08	± 0.2	± 0.1

A.3 Details concerning the experimental setup

The computer system on which the algorithms have been run for conducting the experiments had the following configuration:

- **CPU / RAM:** Intel(R) Core(TM) i7-3770 @ 3.40 GHz, 4 physical cores with 2 hardware threads each; 16 GB RAM.

- **Hard drive:** Solid-State-Drive.
- **Operating system:** Windows 7 64 Bit.

For the calculation of the EMD, a C++ library of the Tree-EMD algorithm [Lin07] has been employed. The fastCMD-algorithm has been implemented in C#.

The two image fusion algorithms and the extraction of the features ScatterCount and ScatterWidth out of a light transport matrix have been implemented in MATLAB.

For the conducted real experiments based on camera images, the following optical components have been employed:

- **Camera:** Monochrome area-scan camera JAI GO-5000M-PGE having a resolution of 2560×2048 and pixels with sizes of 5 μm $\times$ 5 μm.
- **Telecentric lens:** Opto Engineering TC2MHR120-C having a magnification of 0.104.
- **Entocentric lens:** Panasonic H-FS014042.

Bibliography

[Abr13] Milton Abramowitz. *Handbook of mathematical functions.* 9. Dover print. New York, NY: Dover Publ, 2013.

[Ada09a] Francesco Adamo et al. "A low-cost inspection system for online defects assessment in satin glass". In: *Measurement* 42.9 (2009), pp. 1304–1311.

[Ada09b] Francesco Adamo et al. "An online defects inspection system for satin glass based on machine vision". In: *IEEE Instrumentation and Measurement Technology Conference.* 2009, pp. 288–293.

[Ada10] F. Adamo, F. Attivissimo, and A. Di Nisio. "Calibration of an Inspection System for Online Quality Control of Satin Glass". In: *IEEE Transactions on Instrumentation and Measurement* 59.5 (2010), pp. 1035–1046.

[Ade91] Edward H Adelson and James R Bergen. "The Plenoptic Function and the Elements of Early Vision". In: *Computational Models of Visual Processing* (1991), pp. 3–20.

[Agr08] Ilka Agricola and Thomas Friedrich. *Elementary Geometry.* Providence, R.I: American Mathematical Society, 2008.

[Akd15] Bayram Akdemir and Saban Öztürk. "Glass Surface Defects Detection with Wavelet Transforms". In: *International Journal of Materials, Mechanics and Manufacturing* 3.3 (2015), pp. 170–173.

[Ant] Antares Vision. URL: `antaresvisionindustrial.com` (visited on 05/30/2018).

[Ant16] Nicholas Antipa et al. "Single-shot diffuser-encoded light field imaging". In: *IEEE International Conference on Computational Photography*. 2016, pp. 1–11.

[Bau06] Torsten Baumbach et al. "Remote Metrology by Comparative Digital Holography". In: *Applied Optics* 45.5 (2006), pp. 925–934.

[Ber16] Stephan Bergmann et al. "A Phenomenological Approach to Integrating Gaussian Beam Properties and Speckle into a Physically-Based Renderer". In: *Conference on Vision, Modeling and Visualization*. 2016, pp. 179–186.

[Bey15] Jürgen Beyerer, Fernando Puente León, and Christian Frese. *Machine Vision: Automated Visual Inspection: Theory, Practice and Applications*. Springer Berlin Heidelberg, 2015.

[Boh07] Craig F. Bohren and Donald R. Huffman. *Absorption and Scattering of Light by Small Particles*. Wiley-VCH Verlag GmbH, 2007.

[Bot03] T Bothe et al. "Generation and Evaluation of Object Adapted Inverse Patterns for Projection". In: *tm–Technisches Messen* 70.2 (2003), pp. 99–103.

[Cao14] Xuan Cao, Zheng Geng, and Tuotuo Li. "Dictionary-based light field acquisition using sparse camera array". In: *Optics Express* 22.20 (2014), pp. 24081–24095.

[Cha07] Sung-Hyuk Cha. "Comprehensive survey on distance/similarity measures between probability density functions". In: *International Journal of Mathematical Models and Methods in Applied Sciences* 1.4 (2007), pp. 300–307.

[Cha16] Ming Chang et al. "Development of an optical inspection platform for surface defect detection in touch panel glass". In: *International Journal of Optomechatronics* 10.2 (2016), pp. 63–72.

[Chi82] Roland T. Chin and Charles A. Harlow. "Automated visual inspection: A survey". In: *IEEE Transactions on Pattern Analysis and Machine Intelligence* 6 (1982), pp. 557–573.

[Cro84] Franklin C. Crow. "Summed-area tables for texture mapping". In: *ACM SIGGRAPH Computer Graphics* 18.3 (1984), pp. 207–212.

[Dar] Dark Field Technologies. URL: `darkfield.com` (visited on 05/30/2018).

[Des10] Nikunj Desai, Abhinav Singh, and Daniel J. Valentino. "Practical evaluation of image quality in computed radiographic (CR) imaging systems". In: *Medical Imaging 2010: Physics of Medical Imaging*. Vol. 7622. 2010, pp. 1–10.

[Dua04a] Feng Duan, Yaonan Wang, and Huanjun Liu. "A real-time machine vision system for bottle finish inspection". In: *IEEE Conference on Control, Automation, Robotics and Vision*. Vol. 2. 2004, pp. 842–846.

[Dua04b] Feng Duan et al. "Empty bottle inspector based on machine vision". In: *IEEE International Conference on Machine Learning and Cybernetics*. Vol. 6. 2004, pp. 3845–3850.

[Dua07] Feng Duan et al. "A machine vision inspector for beer bottle". In: *Engineering Applications of Artificial Intelligence* 20.7 (2007), pp. 1013–1021.

[Fib] FiberVision GmbH. URL: `www.fibervision.de` (visited on 05/30/2018).

[Gon08] Rafael C. Gonzalez and Richard E. Woods. *Digital image processing*. 3rd ed. Prentice Hall, 2008.

[Gru10] Robin Gruna and Jürgen Beyerer. "On scene-adapted illumination techniques for industrial inspection". In: *IEEE Instrumentation and Measurement Technology Conference*. 2010, pp. 498–503.

[Gün] Optische Prüfsysteme Dr. Günther. URL: `www.optical-inspections.com` (visited on 05/30/2018).

[Han08] Uwe D. Hanebeck and Vesa Klumpp. "Localized Cumulative Distributions and a Multivariate Generalization of the Cramér-von Mises Distance". In: *IEEE International Conference on Multisensor Fusion and Integration for Intelligent Systems*. 2008, pp. 33–39.

[Hap93] Bruce Hapke. *Theory of Reflectance and Emittance Spectroscopy*. 3. Cambridge, England; New York, USA: Cambridge University Press, 1993.

[Har08] Matthias Hartrumpf et al. "Neues Verfahren zur Sichtprüfung transparenter Materialien". In: *Sensorgestützte Sortierung*. 2008, pp. 57–58.

[Har09] Matthias Hartrumpf and Rüdiger Heintz. *Device and method for the classification of transparent components in a material flow*. Patent WO 2009/049594 A1. 2009.

[Hec17] Eugene Hecht. *Optics*. 5th edition. Boston Columbus Indianapolis: Pearson, 2017.

[Hua14] Fu-Chung Huang et al. "Eyeglasses-free display: towards correcting visual aberrations with computational light field displays". In: *ACM Transactions on Graphics* 33.4 (2014), pp. 1–12.

[Hul57] H.C. van de Hulst. *Light Scattering by Small Particles*. Dover Books on Physics. Dover Publications, 1957.

[Ihr10] Ivo Ihrke et al. "Transparent and Specular Object Reconstruction". In: *Computer Graphics Forum* 29.8 (2010), pp. 2400–2426.

[Ind] Industrial Vision Systems. URL: industrialvision.co.uk (visited on 05/30/2018).

[Int96] International Organization for Standardization. *International standard ISO 10110 – Optics and optical instruments - Preparation of drawings for optical elements and systems*. Geneva, Switzerland, 1996.

[Irg11] Stephan Irgenfried, Igor Tchouchenkov, and Heinz Wörn. "CADaVISION: A Simulation Framework for Machine Vision Prototyping". In: *International Conference on Computer Modeling and Simulation*. 2011, pp. 59–67.

[Irg14] Stephan Irgenfried, Frank Dittrich, and Heinz Wörn. "Realization and evaluation of image processing tasks based on synthetic sensor data: 2 use cases". In: *Forum Bildverarbeitung*. 2014, pp. 35–46.

[Irg17] Stephan Irgenfried et al. "Image formation simulation for computer-aided inspection planning of machine vision systems". In: *Automated Visual Inspection and Machine Vision*. 2017, pp. 1–13.

[IRI] IRIS inspection machines. URL: www.iris-im.com (visited on 05/30/2018).

[ISR] ISRA VISION. URL: www.israglassvision.com (visited on 05/30/2018).

[Jak10] Wenzel Jakob. *Mitsuba renderer*. 2010. URL: www.mitsuba-renderer.org (visited on 07/23/2018).

[Jeo11] Yong-Ju Jeon et al. "Detection of scratch defects on slab surface". In: *IEEE International Conference on Control, Automation and Systems*. 2011, pp. 1274–1278.

[Lam] Lambda-X. URL: `www.lambda-x.com` (visited on 05/30/2018).

[Lan11] Douglas Lanman et al. "Polarization fields: dynamic light field display using multi-layer LCDs". In: *ACM Transactions on Graphics* (2011), pp. 186–194.

[Lev96] Marc Levoy and Pat Hanrahan. "Light field rendering". In: *ACM Conference on Computer Graphics and Interactive Techniques*. 1996, pp. 31–42.

[Li14] Di Li, Lie-Quan Liang, and Wu-Jie Zhang. "Defect inspection and extraction of the mobile phone cover glass based on the principal components analysis". In: *The International Journal of Advanced Manufacturing Technology* 73 (2014), pp. 1605–1614.

[Li15] Anding Li et al. "Online optical quality inspection of float glass by a Moiré method". In: *Glass Technology-European Journal of Glass Science and Technology Part A* 56.1 (2015), pp. 21–27.

[Lia07] Chia-Kai Liang, Liu Gene, and Homer H. Chen. "Light Field Acquisition using Programmable Aperture Camera". In: *IEEE International Conference on Image Processing*. 2007, pp. 233–236.

[Lia08] Chia-Kai Liang et al. "Programmable aperture photography: multiplexed light field acquisition". In: *ACM Transactions on Graphics*. 2008, pp. 1–10.

[Lia16] Lie-Quan Liang et al. "Touch screen defect inspection based on sparse representation in low resolution images". In: *Multimedia Tools and Applications* 75.5 (2016), pp. 2655–2666.

[Lin07] Haibin Ling and Kazunori Okada. "An efficient earth mover's distance algorithm for robust histogram comparison". In: *IEEE Transactions on Pattern Analysis and Machine Intelligence* 29.5 (2007), pp. 840–853.

[Lip08] Gabriel Lippmann. "Epreuves reversibles. Photographies integrals". In: *Comptes-Rendus de l'Académie des Sciences* 146 (1908), pp. 446–451.

[Lyt] Lytro light field cameras. URL: `en.wikipedia.org/wiki/Lytro` (visited on 06/28/2018).

[Ma14] Chenguang Ma et al. "Transparent Object Reconstruction via Coded Transport of Intensity". In: *IEEE Conference on Computer Vision and Pattern Recognition*. 2014, pp. 3238–3245.

[Mal03] Elias N. Malamas et al. "A survey on industrial vision systems, applications and tools". In: *Image and Vision Computing* 21.2 (2003), pp. 171–188.

[Meh08] Kurt Mehlhorn and Peter Sanders. *Algorithms and data structures: The basic toolbox*. Springer Science & Business Media, 2008.

[Mel10] Wolfgang Melchert et al. "Automatische Sichtprüfung strukturierter transparenter Materialien". In: *Forum Bildverarbeitung*. 2010, pp. 143–153.

[Mey14] Johannes Meyer. *Visual Inspection of Transparent Objects – Physical Basics, Existing Methods and Novel Ideas*. Tech. rep. IES-2014-04. Karlsruhe Institute of Technology, 2014, pp. 37–47.

[Mey15] Johannes Meyer. *Overview on Machine Vision Methods for Finding Defects in Transparent Objects*. Tech. rep. IES-2015-08. Karlsruhe Institute of Technology, 2015, pp. 103–112.

[Mey16a] J. Meyer, T. Längle, and J. Beyerer. "Acquiring and Processing Light Deflection Maps for Transparent Object Inspection". In: *2nd International Conference on Frontiers of Signal Processing*. 2016, pp. 104–109.

[Mey16b] Johannes Meyer. *Next on Stage: 'MC ViSi' – a Machine Vision Simulation Framework*. Tech. rep. IES-2016-06. Karlsruhe Institute of Technology, 2016, pp. 71–83.

[Mey16c] Johannes Meyer, Thomas Längle, and Jürgen Beyerer. "About acquiring and processing light transport matrices for transparent object inspection". In: *tm-Technisches Messen* (2016), pp. 731–738.

[Mey16d] Johannes Meyer, Thomas Längle, and Jürgen Beyerer. "About the Acquisition and Processing of Ray Deflection Histograms for Transparent Object Inspection". In: *Irish Machine Vision & Image Processing Conference Proceedings*. 2016, pp. 9–16.

[Mey16e] Johannes Meyer, Thomas Längle, and Jürgen Beyerer. "Acquisition and processing of light transport matrices for automated transparent object inspection". In: *Forum Bildverarbeitung*. 2016, pp. 75–86.

[Mey16f] Johannes Meyer et al. "Simulation of an Inverse Schlieren Image Acquisition System for Inspecting Transparent Objects". In: *Electronic Imaging*. 2016, pp. 1–9.

[Mey17a] Johannes Meyer, Thomas Längle, and Jürgen Beyerer. "fastGCVM: A Fast Algorithm for the Computation of the Discrete Generalized Cramér-von Mises Distance". In: *25th International Conference in Central Europe on Computer Graphics, Visualization and Computer Vision*. 2017, pp. 147–152.

[Mey17b] Johannes Meyer, Thomas Längle, and Jürgen Beyerer. "General Cramér-von Mises, a Helpful Ally for Transparent Object Inspection Using Deflection Maps?" In: *Scandinavian Conference on Image Analysis*. 2017, pp. 526–537.

[Mey17c] Johannes Meyer, Thomas Längle, and Jürgen Beyerer. "Towards light transport matrix processing for transparent object inspection". In: *Computing Conference*. July 2017, pp. 244–248.

[Mey18a] Johannes Meyer, Thomas Längle, and Jürgen Beyerer. "Optical realization and calibration of a light field generator". In: *Proceedings of SPIE Optical Systems Design*. Vol. 10693. 2018, pp. 1–10.

[Mey18b] Johannes Meyer et al. "SNR-optimized image fusion for transparent object inspection". In: *Proceedings of SPIE Photonics Europe*. Vol. 10677. 2018, pp. 1–11.

[mih] miho Inspektionssysteme. URL: `miho.de` (visited on 05/30/2018).

[Miy17] Isao Miyagawa and Tetsuya Kinebuchi. "Compressive Inverse Light Transport for Radiometric Compensation in Projection-based Displays". In: *ITE Transactions on Media Technology and Applications* 5.3 (2017), pp. 96–109.

[Moh16] Mahsa Mohammadikaji et al. "A framework for uncertainty propagation in 3D shape measurement using laser triangulation". In: *IEEE International Conference on Instrumentation and Measurement Technology*. 2016, pp. 1–6.

[Moh17] Mahsa Mohammadikaji et al. "Probabilistic surface inference for industrial inspection planning". In: *IEEE Winter Conference on Applications of Computer Vision*. 2017, pp. 1008–1016.

[Mon81] Gaspard Monge. "Mémoire sur la théorie des déblais et des remblais". In: *Histoire de l'Académie Royale des Sciences de Paris* (1781).

[Ng05] Ren Ng et al. "Light field photography with a hand-held plenoptic camera". In: *Computer Science Technical Report CSTR* 2.11 (2005), pp. 1–11.

[Ng06] Ren Ng. *Digital light field photography*. Stanford University California, 2006.

[Nin] Nintendo 3DS. URL: https://www.nintendo.co.uk/Nintendo-3DS-Family/Nintendo-3DS-Family-94560.html (visited on 06/28/2018).

[Nür16] Thomas Nürnberg, Christian Zimmermann, and Fernando Puente León. "Simulationsgestützte Optimierung einer Computational-Kamera zur dichten Tiefenschätzung". In: *tm - Technisches Messen* 83.9 (2016), pp. 511–520.

[OPO] OPOS. URL: opos.de (visited on 05/30/2018).

[Ost02] Wolfgang Osten, Torsten Baumbach, and Werner Jüptner. "Comparative Digital Holography". In: *Optics Letters* 27.20 (2002), pp. 1764–1766.

[OTo10] Matthew O'Toole and Kiriakos N. Kutulakos. "Optical computing for fast light transport analysis". In: *ACM Transactions on Graphics* 29.6 (2010), pp. 164–174.

[OTo14] Matthew O'Toole, John Mather, and Kiriakos Kutulakos. "3d shape and indirect appearance by structured light transport". In: *IEEE Conference on Computer Vision and Pattern Recognition*. 2014, pp. 3246–3253.

[Pap02] Athanasios Papoulis and S. Unnikrishna Pillai. *Probability, random variables, and stochastic processes*. 4th. McGraw-Hill, 2002.

[Pee09] Pieter Peers et al. "Compressive light transport sensing". In: *ACM Transactions on Graphics* 28 (2009), pp. 1–18.

[Pel89] Shmuel Peleg, Michael Werman, and Hillel Rom. "A unified approach to the change of resolution: Space and gray-level". In: *IEEE Transactions on Pattern Analysis and Machine Intelligence* 11.7 (1989), pp. 739–742.

[Pha16] Matt Pharr, Wenzel Jakob, and Greg Humphreys. *Physically based rendering: From theory to implementation*. Morgan Kaufmann, 2016.

[Pla01] Ben C. Platt and Roland Shack. "History and principles of Shack-Hartmann wavefront sensing". In: *Journal of Refractive Surgery* 17.5 (2001), pp. 573–577.

[Pös12] Andreas Pösch, Taras Vynnyk, and Eduard Reithmeier. "Using inverse fringe projection to speed up the detection of local and global geometry defects on free-form surfaces". In: *SPIE Proceedings on Image Reconstruction from Incomplete Data VII*. 2012, pp. 8500–8507.

[Ray] Raytrix light field cameras. URL: raytrix.de (visited on 05/29/2018).

[REA] REALEYES. URL: real-eyes.eu/en/ (visited on 06/28/2018).

[Ros73] Albert Rose. *Vision: human and electronic*. Optical physics and engineering. New York: Plenum Press, 1973.

[Rub00] Yossi Rubner, Carlo Tomasi, and Leonidas J. Guibas. "The earth mover's distance as a metric for image retrieval". In: *International Journal of Computer Vision* 40.2 (2000), pp. 99–121.

[Sat09] S. Satorres Martínez et al. "A sensor planning system for automated headlamp lens inspection". In: *Expert Systems with Applications* 36.5 (2009), pp. 8768–8777.

[Sat12] S. Satorres Martínez et al. "A machine vision system for defect characterization on transparent parts with non-plane surfaces". In: *Machine Vision and Applications* 23 (2012), pp. 1–13.

[Sch07] Y.Y. Schechner, S.K. Nayar, and P.N. Belhumeur. "Multiplexing for Optimal Lighting". In: *IEEE Transactions on Pattern Analysis and Machine Intelligence* 29.8 (2007), pp. 1339–1354.

[Sch73] James W. Schoonard and John D. Gould. "Field of view and target uncertainty in visual search and inspection". In: *Human Factors* 15.1 (1973), pp. 33–42.

[See] SeeReal Technologies. URL: www.seereal.com (visited on 06/28/2018).

[Set01] G. S. Settles. *Schlieren and Shadowgraph Techniques Visualizing Phenomena in Transparent Media*. Berlin, Heidelberg: Springer, 2001.

[Sha15] Wenjie Shao, Hao Sheng, and Chao Li. "Segment-Based Depth Estimation in Light Field Using Graph Cut". In: *International Conference on Knowledge Science, Engineering and Management*. Springer, 2015, pp. 248–259.

[She] Shelton Vision. URL: `www.sheltonvision.co.uk` (visited on 05/30/2018).

[Sof] Softsolution GmbH Glass IQ. URL: `www.glass-iq.com` (visited on 05/30/2018).

[Sou] César Souza. *The Accord.NET Framework*. URL: `accord-framework.net` (visited on 07/29/2018).

[Sou15] Daniel Soukup, Svorad Štolc, and Reinhold Huber-Mörk. "Analysis of optically variable devices using a photometric light-field approach". In: *IS&T/SPIE Electronic Imaging*. 2015, pp. 1–9.

[Što14] Svorad Štolc et al. "Depth and all-in-focus images obtained by multi-line-scan light-field approach". In: *IS&T/SPIE Electronic Imaging*. 2014, pp. 1–16.

[Sud13] Prasad Sudhakar et al. "Compressive schlieren deflectometry". In: *IEEE International Conference on Acoustics, Speech and Signal Processing*. 2013, pp. 5999–6003.

[Sud15] Prasad Sudhakar et al. "Compressive imaging and characterization of sparse light deflection maps". In: *SIAM Journal on Imaging Sciences* 8.3 (2015), pp. 1824–1856.

[Syn] Synopsys LightTools. URL: `www.synopsys.com` (visited on 05/30/2018).

[Tak16] Yuichiro Takeuchi, Shunichi Suwa, and Kunihiko Nagamine. "AnyLight: Programmable Ambient Illumination via Computational Light Fields". In: *ACM International Conference on Interactive Surfaces and Spaces*. 2016, pp. 39–48.

[Tos14] Ivana Tosic and Kathrin Berkner. "Light field scale-depth space transform for dense depth estimation". In: *IEEE Conference on Computer Vision and Pattern Recognition Workshops*. 2014, pp. 435–442.

[Tox13] C. Toxqui-Quitl et al. "Bottle inspector based on machine vision". In: *Image Processing: Machine Vision Applications*. 2013, pp. 8661–8670.

[Vip] Viprotron GmbH. URL: `www.viprotron.de` (visited on 05/30/2018).

[Vis] VisuMotion. URL: `www.visumotion.com` (visited on 06/28/2018).

[Wan00] Jun Wang and Anand K. Asundi. "A computer vision system for wineglass defect inspection via Gabor-filter-based texture features". In: *Information Sciences* 127.3 (2000), pp. 157–171.

[Wan09] Jiaping Wang et al. "Kernel Nyström method for light transport". In: *Transactions on Graphics*. Vol. 28. 2009, pp. 29–38.

[Wan14] Sven Wanner and Bastian Goldlüecke. "Variational Light Field Analysis for Disparity Estimation and Super-Resolution". In: *IEEE Transactions on Pattern Analysis and Machine Intelligence* 36.3 (2014), pp. 606–619.

[Wer07] Stefan Werling and Jürgen Beyerer. "Inspection of Specular Surfaces with Inverse Patterns". In: *tm–Technisches Messen* 74.4 (2007), pp. 217–223.

[Wet11a] Gordon Wetzstein et al. "Layered 3D: tomographic image synthesis for attenuation-based light field and high dynamic range displays". In: *ACM Transactions on Graphics* 30.4 (2011), pp. 95–105.

[Wet11b] Gordon Wetzstein et al. "Refractive shape from light field distortion". In: *IEEE International Conference on Computer Vision*. 2011, pp. 1180–1186.

[Wet12] Gordon Wetzstein et al. "Tensor displays: compressive light field synthesis using multilayer displays with directional backlighting". In: *ACM Transactions on Graphics* 31.4 (2012), pp. 1–11.

[Yan02] Jason C. Yang et al. "A real-time distributed light field camera". In: *Rendering Techniques* 2002 (2002), pp. 77–86.

[Yao16] Li Yao, Yunjian Liu, and Weixin Xu. "Real-time virtual view synthesis using light field". In: *EURASIP Journal on Image and Video Processing* 2016.1 (2016), pp. 1–10.

[Yu11] Wenyong Yu. "On-line defects inspection of floating glass by variable LED raster". In: *Remote Sensing Image Processing, Geographic Information Systems, and Other Applications*. Ed. by Jianguo Liu et al. 2011, pp. 1–7.

[Zem] Zemax. URL: `www.zemax.com` (visited on 06/28/2018).

List of publications

[Mey14] Johannes Meyer. *Visual Inspection of Transparent Objects – Physical Basics, Existing Methods and Novel Ideas*. Tech. rep. IES-2014-04. Karlsruhe Institute of Technology, 2014, pp. 37–47.

[Mey15] Johannes Meyer. *Overview on Machine Vision Methods for Finding Defects in Transparent Objects*. Tech. rep. IES-2015-08. Karlsruhe Institute of Technology, 2015, pp. 103–112.

[Mey16a] J. Meyer, T. Längle, and J. Beyerer. "Acquiring and Processing Light Deflection Maps for Transparent Object Inspection". In: *2nd International Conference on Frontiers of Signal Processing*. 2016, pp. 104–109.

[Mey16b] Johannes Meyer. *Next on Stage: 'MC ViSi' – a Machine Vision Simulation Framework*. Tech. rep. IES-2016-06. Karlsruhe Institute of Technology, 2016, pp. 71–83.

[Mey16c] Johannes Meyer, Thomas Längle, and Jürgen Beyerer. "About acquiring and processing light transport matrices for transparent object inspection". In: *tm-Technisches Messen* (2016), pp. 731–738.

[Mey16d] Johannes Meyer, Thomas Längle, and Jürgen Beyerer. "About the Acquisition and Processing of Ray Deflection Histograms for Transparent Object Inspection". In: *Irish Machine Vision & Image Processing Conference Proceedings*. 2016, pp. 9–16.

[Mey16e] Johannes Meyer, Thomas Längle, and Jürgen Beyerer. "Acquisition and processing of light transport matrices for automated transparent object inspection". In: *Forum Bildverarbeitung*. 2016, pp. 75–86.

[Mey16f] Johannes Meyer et al. "Simulation of an Inverse Schlieren Image Acquisition System for Inspecting Transparent Objects". In: *Electronic Imaging*. 2016, pp. 1–9.

[Mey17a] Johannes Meyer, Thomas Längle, and Jürgen Beyerer. "fastGCVM: A Fast Algorithm for the Computation of the Discrete Generalized Cramér-von Mises Distance". In: *25th International Conference in Central Europe on Computer Graphics, Visualization and Computer Vision*. 2017, pp. 147–152.

[Mey17b] Johannes Meyer, Thomas Längle, and Jürgen Beyerer. "General Cramér-von Mises, a Helpful Ally for Transparent Object Inspection Using Deflection Maps?" In: *Scandinavian Conference on Image Analysis*. 2017, pp. 526–537.

[Mey17c] Johannes Meyer, Thomas Längle, and Jürgen Beyerer. "Towards light transport matrix processing for transparent object inspection". In: *Computing Conference*. July 2017, pp. 244–248.

[Mey18a] Johannes Meyer, Thomas Längle, and Jürgen Beyerer. "Optical realization and calibration of a light field generator". In: *Proceedings of SPIE Optical Systems Design*. Vol. 10693. 2018, pp. 1–10.

[Mey18b] Johannes Meyer et al. "SNR-optimized image fusion for transparent object inspection". In: *Proceedings of SPIE Photonics Europe*. Vol. 10677. 2018, pp. 1–11.

SCHRIFTENREIHE AUTOMATISCHE SICHTPRÜFUNG UND BILDVERARBEITUNG
(ISSN 1866-5934)

Herausgeber: Prof. Dr.-Ing. habil. Jürgen Beyerer

Band 1 **JONATHAN BALZER**
Regularisierung des Deflektometrieproblems Grundlagen und Anwendung. 2008
ISBN 978-3-86644-230-6

Band 2 **IOANA GHEŢA**
Fusion multivariater Bildserien am Beispiel eines Kamera-Arrays. 2011
ISBN 978-3-86644-684-7

Band 3 **STEFAN BRUNO WERLING**
Deflektometrie zur automatischen Sichtprüfung und Rekonstruktion spiegelnder Oberflächen. 2011
ISBN 978-3-86644-687-8

Band 4 **JAN WASSENBERG**
Efficient Algorithms for Large-Scale Image Analysis. 2012
ISBN 978-3-86644-786-8

Band 5 **MARTIN GRAFMÜLLER**
Verfahrensfortschritte in der robusten Echtzeiterkennung von Schriftzeichen. 2013
ISBN 978-3-86644-979-4

Band 6 **JÜRGEN BRAUER**
Human Pose Estimation with Implicit Shape Models. 2014
ISBN 978-3-7315-0184-8

Band 7 **MARKUS MÜLLER**
Szeneninterpretation unter Verwendung multimodaler Sensorik und Salienzmaßen. 2017
ISBN 978-3-7315-0240-1

Band 8 **ROBIN GRUNA**
Beleuchtungsverfahren zur problemspezifischen Bildgewinnung für die automatische Sichtprüfung. 2015
ISBN 978-3-7315-0313-2

Band 9 **THOMAS STEPHAN**
Beitrag zur Unterwasserbildrestauration. 2017
ISBN 978-3-7315-0579-2

Die Bände sind unter www.ksp.kit.edu als PDF frei verfügbar oder als Druckausgabe bestellbar.

Band 10 **JAN-PHILIP JARVIS**
A Contribution to Active Infrared Laser Spectroscopy for Remote Substance Detection. 2017
ISBN 978-3-7315-0725-3

Band 11 **MIRO TAPHANEL**
Chromatisch konfokale Triangulation – Hochgeschwindigkeits 3D-Sensorik auf Basis der Wellenlängenschätzung mit optimierten Filtern. 2018
ISBN 978-3-7315-0646-1

Band 12 **SEBASTIAN HÖFER**
Untersuchung diffus spiegelnder Oberflächen mittels Infrarotdeflektometrie. 2017
ISBN 978-3-7315-0711-6

Band 13 **MATTHIAS RICHTER**
Über lernende optische Inspektion am Beispiel der Schüttgutsortierung. 2018
ISBN 978-3-7315-0842-7

Band 14 **MATHIAS ZIEBARTH**
Wahrnehmungsgrenzen kleiner Verformungen auf spiegelnden Oberflächen. 2019
ISBN 978-3-7315-0890-8

Band 15 **JOHANNES MEYER**
Light Field Methods for the Visual Inspection of Transparent Objects. 2019
ISBN 978-3-7315-0912-7